Comparative Models for Electrical Load Forecasting

Comparative Models for Electrical Load Forecasting

Edited by

D. W. Bunn
London Graduate School of Business Studies

and

E. D. Farmer
Imperial College of Science and Technology

JOHN WILEY & SONS
Chichester · New York · Brisbane · Toronto · Singapore

Library of Congress Cataloging in Publication Data:
Main entry under title:

Comparative models for electrical load forecasting.

Includes bibliographical references and indexes.
1. Electric power-plants—Load—Mathematical models.
I. Bunn, Derek W. II. Farmer, E. D. (E. Dillon)
TK1191.C62 1985 363.6′22′0724 84–20873
ISBN 0 471 90635 2

British Library Cataloging in Publication Data:

Comparative models for electrical load forecasting.
1. Electric power system—Load dispatching
I. Bunn, Derek W. II. Farmer, E. D.
621.37 TK1005
ISBN 0 471 90635 2

Printed in Northern Ireland at The Universities Press (Belfast) Ltd.

Contents

Preface

A large number of papers concerned with forecasting the loads on public utility systems has appeared over the past twenty years or so in journals and conference proceedings. However, from a practical perspective, these publications have a limited usefulness. The very nature of professional journals and conferences encourages the presentation of novel, often speculative and insufficiently tested, methods which tend to emphasize theoretical rather than implementational aspects. Thus, despite this literature, there is very little published material relating to how short-term forecasting has actually been undertaken and is being developed in public utility organizations.

In this volume, it has been the editors' intention to encourage some of the scientists working in this area to describe their operational procedures and the sorts of developmental problems which have had to be overcome. As a study in *comparative research* into forecasting practice, this work is of value in documenting the various applied approaches which have been taken to an essentially similar problem faced by all power utilities.

Short-term forecasting in this context will be taken to include the spectrum of methodology developed to deal with lead times of up to approximately twenty-four hours. In the very short-term, state estimation techniques are designed to validate current system measurements and estimate the load. For lead times of a few minutes to a few hours, automatic adaptive forecasting methods are required to support procedures for scheduling the secure output of generators. The constraints of on-line operation usually necessitate a univariate method of forecasting at this level. For lead times of several hours, multivariate forecasting procedures are commonly used to relate the demand to changes in the weather. For this category, the forecasting problem in the electricity industry is similar to that in other utilities such as gas and water. It is hoped, therefore, that this book will be of interest to a wide readership in forecasting, beyond those specializing in the electricity and energy sectors.

The volume is divided into three parts. Part One is editorial, comprising a chapter introducing the general economic and operational decision-making context that short-term load forecasting supports and a second chapter reviewing the published literature on short-term load forecasting methodology. Part Two consists of six contributed chapters which provide comparative insights on a range of model developments in load forecasting. A variety of time-series models are presented, including Box–Jenkins, Kalman, Wiener, Spectral, and Double Exponential Smoothing, together with multivariate models for utilizing weather data. Implementational issues concerned with the on-line development of these forcasting models, how they deal with special situations, and the combination of on-line univariate predictors with off-line multivariate forecasts are discussed within several modelling contexts. Part Three looks for further insights from fields adjacent to that of daily electric load forecasting. The chapter by Ramanathan, Granger, and Engle develops an econometric perspective on the subject whilst that of Irving and Sterling looks at the very short-term implications of state-estimation for data validation. The final two chapters develop versions of the Box–Jenkins model identification procedures for the short-term forecasting of gas and water demands.

We, as editors, are grateful for the cooperation of our contributors and publisher in the realization of this project and for the support and facilities provided by the several institutions with which we have been affiliated during its three-year evolution, viz. The Central Electricity Generating Board, Oxford and Stanford Universities, Imperial College, and the London Business School.

Derek Bunn
Dillon Farmer
London, 1984

List of Contributors

Gary Ackerman
Systems Control Inc
1801 Page Mill Road
P.O. Box 10025
Palo Alto, Ca. 94303
USA

A. B. Baker
CEGB Operations Division
Park Street
London SE1
England

A. Bargiela
Department of Engineering
University of Durham
South Road
Durham DH1 3LE
England

D. W. Bunn
London Business School
Sussex Place
Regent's Park
London NW1 4SA
England

Umberto di Caprio
ENEL-DSR
Centro Ricerce di Automatica
Via A. Volta 1
20093 Cologno Monzese
Milano
Italy

Robert Engle
Department of Economics
University of California San Diego
La Jolla, Ca. 92093
USA

E. D. Farmer
Department of Electrical Engineering
Imperial College
London SW7
England

Roberto Genesio
Politecnico di Torina
Dipartimento di Automatica e Informatica
Cors Ducu degli Abruzzi 24
10129 Torino
Italy

Clive Granger
Department of Economics
University of California San Diego
La Jolla, Ca. 92093
USA

P. C. Gupta
EPRI
3412 Hillview Avenue
P.O. Box 10412
Palo Alto, Ca. 94303
USA

M. R. Irving
Department of Engineering
University of Durham
South Road
Durham DH1 3LE
England

W. D. Laing
CERL
Lelvin Avenue
Leatherhead
Surrey KT22 7SE
England

J. L. Piggott
British Gas Corporation
London Research Station
Michael Road
London SW6 2AD
England

Sergio Pozzi
ENEL-DSR
Centro Ricerce di Automatica
Via A. Volta 1
20093 Cologno Monzese
Milano
Italy

Ramu Ramanathan
Department of Economics
University of California San Diego
La Jolla, Ca. 92093
USA

Dean A. Schiffman
San Diego Gas and Electric Company
San Diego, Ca. 92112
USA

Alan M. Schneider
Department of Applied Mechanics and Engineering Services
University of California San Diego
La Jolla, Ca. 92093
USA

M. J. H. Sterling
Department of Engineering
University of Durham
South Road
Durham DH1 3LE
England

Tadashi Takenawa
Technical Research and Development Institute
Japan Defense Agency
Tokyo
Japan

Antonia Vicino
Politechnico di Torina
Dipartimento di Automatica e Informatica
Corsa degli Abruzzi 24
10129 Torino
Italy

PART ONE: INTRODUCTION

Comparative Models for Electrical Load Forecasting
Edited by D. W. Bunn and E. D. Farmer

CHAPTER 1

Economic and Operational Context of Electric Load Prediction

D. W. Bunn and E. D. Farmer

1 INTRODUCTION

Load forecasting occupies a central role in the operation and planning of electric power systems. The lead times range from a few minutes ahead for the economic loading of power plant to over forty years for the economic planning of new generating capacity and transmission networks. A large variety of techniques have been investigated and a recent survey[1] lists over seventy publications on load forecasting philosophy and methodology. The techniques may be broadly divided into three areas of application; these are long-term econometric forecasts for system planning, medium-term forecasts for the scheduling of fuel supplies and maintenance programmes, and short-term predictions for the day-to-day operation and scheduling of the power system. It is the short-term operational context that is the subject of the present volume.

The purpose of the present chapter is to review the economic and operational implications of short-term load forecasts and to outline system requirements in relation to differing predictive applications. In this context, the forecasting systems may be broadly divided into off-line and on-line predictors. The former are principally applied to the scheduling of steam-driven generating plant from a few hours to a few days ahead, whereas the on-line forecasts are used for the minute-to-minute operation of the power system and for the economic loading of the generating units. In the past, off-line methods have employed regression models in which the load is related to exogenous variables, such as meteorological factors.[2] In contrast, on-line forecasting procedures adapt to the minute-to-minute evolution of the most recent telemetered demand values, by the use of time-series techniques.[3] More recently, procedures that utilize both exogenous variables and autoregressive models have been applied. Examples of both off-line methods[4,5] and on-line techniques[6,7] are presented in this volume. A third approach is that of interactive prediction, in which an automatic system allows the optional entry of off-line information into an on-line control and prediction facility.[8] This method seeks to combine the responsiveness of on-line systems to changing load patterns with the longer-term accuracy derived from meteorological factors.

The economic and operational implications of predictive plant scheduling and economic loading are described under separate headings.

2 PLANT SCHEDULING

In the operation of a power system, in order to meet the variation in demand throughout the day and from day to day, it is necessary to plan in advance the start-up and shut-down schedules of the power plant, especially the large steam-driven generating units. This process, which is effected from several hours to a day or so ahead, is often termed unit commitment; it ensures that there is sufficient operational generating capacity to meet the variable load, with a specified reliability, at each moment of the day. The time-scales of the required forecasts are determined by the start-up period of steam plant, the economic effects of start-up costs, dynamical operating constraints, and the need to make appropriate staffing and technical preparations for the running of plant.

The scheduling process involves a complex calculation, in which the expected cost of meeting the demand is minimized subject to a set of static and dynamic plant limits and system security constraints. In the past, this optimization was performed by empirical methods, in which the system operator minimized generating costs at the peaks and troughs of the load curve, and interpolated the schedule between these salient points. Such methods required predictions of the maxima and minima of the load during the critical costing periods of the day. However, in recent years computer-based scheduling systems, which minimize the cost of following the detailed demand profile, have been adopted. Methods for effecting these calculations, including integer programming,[9,10] Lagrangian relaxation,[11] and dynamic programming,[12,13] have been developed.

For the purpose of following a variable load with steam generation, it is important to maintain an optimum economic balance between the costs of starting-up and shutting-down plant and that of running excess part-loaded generation between periods of peak demand. This balance depends upon the magnitudes of the start-up costs and the 'no-load' running costs. The operational cost function of a typical steam-driven generating unit is illustrated in Figure 1.1. Assuming an approximately linear cost curve, the function may be described by two parameters, namely the total cost c_A, per unit of energy (MWh) at the rated output, and the incremental cost c_B for a unit change in output. The cost per hour of holding one unit of excess generating capacity is equal to the 'no-load' cost $c_A - c_B$, where c_A and c_B

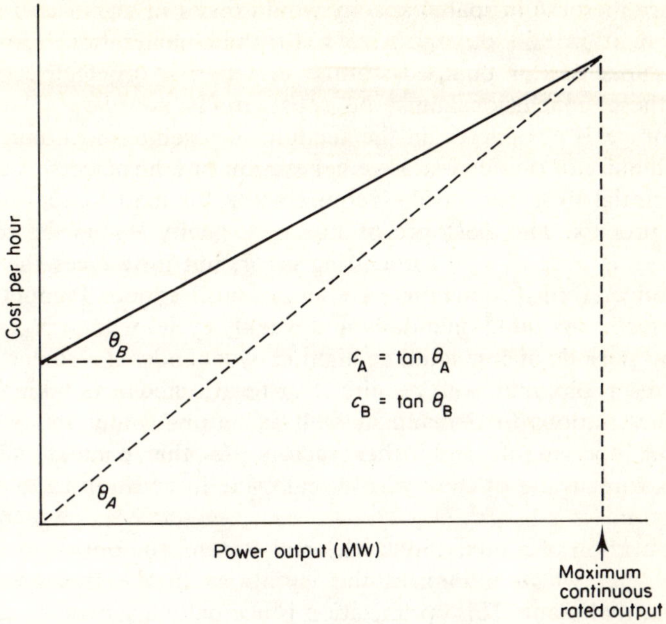

FIGURE 1.1 Linearized cost curve of steam-driven power unit

are the cost parameters associated with the marginal generating plant. This 'no-load' cost ranges from 10% to 20% of the cost of production of one energy unit. The cost of start-up, per unit of generating capacity, ranges from about 2 hours to 20 hours worth of no-load cost, depending upon the design of the plant and the length of previous shut-down period. Clearly, the scheduling procedure, in order to optimize the balance between part-loading and shutting-down plant, requires predictions up to about 24 hours ahead of the event. Such longer term schedules may be progressively refined, by the adjustment of start and stop times, as more accurate load forecast information becomes available. Short-term revision of schedules may be implemented down to lead-times equal to the minimum period for start-up, i.e. a few hours ahead.

In considering the cost penalty due to prediction error in the scheduling of a predominantly thermal power system, it may be noted that each unit of the standard deviation of the error requires an approximately equal increase in the scheduled spare capacity to maintain a similar level of system reliability. In this context, the cost of one unit of spare generating capacity is the 'no-load' cost $c_A - c_B$ of the marginal plant. In the UK, at present prices (1984), an increase of 1% in forecasting error would be associated with an increase in operating costs of about £10 million per year. Failure to provide the necessary increase in spare capacity would result in higher cost penalties by enhanced utilization of expensive gas-turbine generation. However, if significant capacities of pumped-storage or other hydroelectric plant are available, these conclusions must be substantially modified. The case of pumped-storage is considered in the section on reserve allocation.

The availability of pumped-storage generation or hydroelectric generation has a substantial influence on the requirements for load forecasting in the scheduling process. The existence of storage capacity lowers the sensitivity of the operating cost to load forecasting error, but introduces the requirement for longer-term predictions of energy consumption. Pumped-storage plant is normally operated on a daily and weekly cycle, with pumping taking place during periods of low load, at night or weekends. Hydroelectric plant with large reservoirs may well require long-term scheduling taking account of seasonal variations in demand as well as natural water inflows due to precipitation, snowment, and other factors. In this context, scheduling necessitates forecasting of the hydrological cycle in addition to longer term load prediction.

In the operation of a mixed hydrothermal system, the optimum economic schedule is that which minimizes the variations in the fuel costs of the marginal thermal plant. This optimization is normally achieved by the use of mathematical programming[16] or by the method of Lagrangian relaxation.[17] The most uncertain of the constraint limits, in these calculations, are those that derive from predicted energy demands and forecasted water inflows into reservoirs. The sensitivity of overall operating costs to forecasting error is a complex function of the plant mix, the storage capacities, and the demand profile; this must be evaluated in relation to specific system characteristics.

3 LOAD DISPATCHING AND SECURITY ASSESSMENT

In power system operation, load dispatching refers to the minute-to-minute economic allocation of loads to the generating units, to meet the varying demand at minimum cost, whilst ensuring the requisite degree of system security. The calculation necessitates the minimization of a cost function subject to a large number of operating constraints that derive principally from generating plant characteristics and transmission system limits. Many methods for effecting this calculation have been developed, the principal techniques being mathematical programming and Lagrangian methods.[18,19]

In considering the predictive requirements for load dispatching, account must be taken of the restrictions on the rates of change of outputs of power plant and the security constraints relating to power flows along the transmission network. Load dispatching is, essentially, an on-line activity and is normally associated with an on-line prediction system. For this reason, dispatch predictions have centred on time series methods that utilize the on-line telemetered load data that is automatically logged by SCADA or control room computing systems. Techniques employed have included spectra analysis,[20–22] double exponential,[6] and Box–Jenkins methods.[23,24] For the purpose of dispatching, detailed predictions of the load curve up to about 30 minutes ahead, that are revised automatically every minute or so, normally suffice.

A further requirement for the secure economic loading of plant is the provision of short-term forecasts of the demands at the principal supply points of the power system network. Such estimates are necessary in order to calculate the flows on the transmission circuits, associated with any pattern of generation, and so enforce transmission security constraints. The prediction of supply point demands is usually done empirically. One such method utilizes a set of stored ratios that take account of the daily, weakly, and seasonal variations of the ratio between each supply point load and the total demand. These forecasts are also utilized to make predictive assessments of the security of the supply system.

The effect of prediction errors on a load dispatching system may be evaluated in terms of system economic and technical performance. The increase in operating cost due to a dispatching error is not usually substantial, unless there is a significant difference between the incremental and decremental costs of the marginal generating plant. On the other hand, large errors in dispatched loads result in excessive and possibly erratic control activity by the load-following plant and this, in turn, can impose a burden on boiler and turbine control systems resulting in high 'dynamic' costs and increased wear-and-tear. The availability of a substantial tranche of hydroelectric generation reduces the overall sensitivity of dispatch to forecasting errors, but the technical effectiveness of the dispatching system remains an important consideration in evaluating the effect of prediction error.

A further application of short-term forecasts is that of predictive assessment of the security of the power system. The assessment often takes the form of a contingency analysis, in which the effect of the loss of each generator or each transmission circuit is evaluated in turn. Predicted potential overloads are displayed to the system operator, who modifies the generation pattern or network configuration to remove the effect of the offending contingency. Due to the somewhat imprecise and variable nature of transmission limits, the security assessment does not exhibit the same sensitivity to forecasting error as do the economic and dynamic performance of the scheduling and dispatching procedures.

4 STATISTICAL ALLOCATION OF RESERVE CAPACITY

To cater for the unpredictable component of power system load, and the variations in power plant capabilities, it is necessary to schedule additional 'reserve' generating capacity. This reserve avoids excessive utilization of high-cost standby plant and contains the frequency of supply interruptions at acceptable levels. The optimum magnitude of the reserve capacity, on a large power system, is dependent upon the forecasting error of predictions with a lead time equal to the shortest period required to bring the next marginal generator on to load. To illustrate the dependence of reserve level on forecasting error, a thermal power system with significant capacities of pumped-storage and gas-turbine plant will be considered.

If σ_s denotes the standard deviation of the load prediction error, and $\gamma_s\sigma_s$, $\gamma_p\sigma_s$, and $\gamma_G\sigma_s$ are the scheduled steam reserve capacity, the pumped-storage reserve, and the gas-turbine capacity, respectively, then the total reserve $(\gamma_s+\gamma_p+\gamma_G)\sigma_s$ is used to service load deviations from the predicted value. If c_B, c_P, and c_G are the incremental costs of the marginal steam plant, of pumped storage generation, and of gas-turbine units, respectively, it may be assumed that the costs satisfy the inequalities:

$$c_B < c_P < c_G. \tag{1.1}$$

In this circumstance, load shortfalls below the predicted value will result in the de-loading of steam plant, whereas demand above the forecasted level will be dispatched, in order of merit, to steam, pumped-storage, and gas-turbines, respectively. It may be shown[25] that the statistical expectation of the total operating cost above the scheduled cost takes the form:

$$c = \gamma_s\sigma_s(c_A - c_B) + \sigma_s(c_P - c_B)G(\gamma_s) + \sigma_s(c_G - c_P)G(\gamma_s + \gamma_p) \tag{1.2}$$

where $G(\gamma)$ is the reserve utilization function given by:

$$G(\gamma) = \int_{\gamma}^{\infty} (x-\gamma)Z(x)\,\mathrm{d}x \tag{1.3}$$

$Z(x)$ being the probability density function of the prediction error, normalized to the standard deviation σ_s. The optimum allocation of steam reserve is obtained by minimizing the expected cost c, in equation (1.2). Equating the derivative to zero yields:

$$(c_P - c_B)Q(\gamma_s) + (c_G - c_P)Q(\gamma_s + \gamma_p) = c_A - c_B \tag{1.4}$$

where $Q(\gamma)$ is the cumulative probability distribution of the load prediction error. Equation (1.4) determines the optimum steam reserve $\gamma_s\sigma_s$ by balancing the costs of invoking pumped-storage and gas-turbine generation against the cost of holding spare generating capacity on the marginal steam plant.

The above analysis demonstrates that the optimal scheduling of generation and reserve capacity requires a knowledge, not only of the expected load curve, but also of the distribution of the forecasting errors. A recent study[26] has shown that the statistical allocation of reserve may be incorporated within the framework of conventional linear programming load dispatching algorithms. The development of an interactive load prediction facility that includes the on-line estimation of error statistics, for this purpose, has been initiated.[27]

5 CONCLUSIONS

The role of load forecasting in power system operation has been briefly reviewed. Predictions are necessary for load dispatching, reserve allocation, security assessment, and short and long term scheduling of generating plant. Predictive requirements include not only the expected demand curve over the defined time horizons, but also the individual loads at the main supply points of the transmission network, and the statistical distribution of the forecasting error as it varies throughout the day and time of year.

The quality of control of the power system, and the economy of operation, are highly sensitive to forecasting error. In consequence, it is expedient to develop forecasting techniques to the stage where the magnitudes of the errors have attained irreducible proportions. In this respect, economic penalties stem principally from errors in the scheduling process, whereas dispatching errors lead to problems in the control of system frequency, exchanges between utilities, and the control of individual boiler and turbine systems.

Load forecasting facilities may be broadly classified as off-line for scheduling and operational planning and on-line for system control and dispatching. In the on-line facilities, an important determining factor in overall performance is the quality of the input data that derives from the telemetered information that is logged by the control room computing system. Such data is often subject to both steady-state random errors and occasional large intermittent errors. Sophisticated methods for the detection and correction of such errors have been developed and implemented; one such scheme for data validation and state-estimation is the subject of a chapter of this volume.[28]

Load forecasting plays a dominant part in the economic optimization and secure operation of electrical power systems. The high cost and limited resources of primary energy inputs, together with the development of computer and control technologies, provide incentive for the further development of techniques for load prediction for system operation.

REFERENCES

1. IEEE Committee (1980). 'Load forecasting bibliography.' *IEEE Transactions*, **PAS-99,** No. 1.
2. M. Davies. *The relationship between weather and electricity demand.*

3. E. D. Farmer (1964). 'A method of prediction for non-stationary processes and its applications to load forecasting.' *Automation and Remote Control*, Butterworths.
4. A. B. Baker. Chapter 5 of this volume.
5. A. M. Schneider, T. Takenawa, and D. A. Schiffman. Chapter 7 of this volume.
6. W. D. Laing. Chapter 6 of this volume.
7. U. di Caprio, R. Genesio, S. Pozzi, and A. Vicino. Chapter 8 of this volume.
8. E. D. Farmer, W. D. Laing, A. M. Adatia, A. B. Baker, and D. W. Bunn (1981). 'The development and implementation of an interactive demand validation and prediction facility.' *Proc. Seventh Power Systems Computations Conf., 1981.*
9. A. Ohuchi and I. Kaji (1975). 'A branch and bound algorithm for the start-up and shut-down problem of thermal generation.' *Elect. Eng in Japan,* **95**.
10. T. S. Dillon, K. W. Edwin, H. D. Kochs, and R. J. Taud (1978). 'Integer programming approach to the problem of unit commitment.' *IEEE Transactions,* **PAS-97**.
11. D. P. Bertsekas, G. S. Lauer, N. R. Sandell, and T. A. Posbergh (1983). 'Optimal short-term scheduling of large scale power systems.' *IEEE Transactions on Automatic Control,* **AC-28,** No. 1.
12. R. D. Machate (1978). *Kurzfristige Kraftwerkseinsatzoptimierung mit Hilfe Dynamischer Optimierung*, Energy Research Society Report, University of Aachen.
13. K. W. Edwin, R. D. Machate, and R. J. Taud (1980). 'Comparative evaluation of unit commitment approaches.' *IEE International Conf. Power System Monitoring and Control, 1980.*
14. T. N. Satra and S. A. Khaparde (1978). 'An application of a direct method to the optimal scheduling of hydrothermal systems.' *IEEE Transactions,* **PAS-97**.
15. J. A. Muchstadt and S. A. Koenig (1977). 'An application of Lagrangian relaxation to scheduling in power systems.' *Operational Research,* **25**(3).
16. M. E. El-Hawary and G. S. Christensen (1979). *Optimal economic operation of electric power systems*, Academic Press, New York.
17. M. V. Pereira (1982). 'Economic dispatch of hydrothermal systems.' *IEEE Transactions,* **PAS-101**.
18. H. H. Happ (1977). 'Optimal power dispatch—a comprehensive survey.' *IEEE Transactions,* **PAS-96**.
19. D. W. Wells (1968). 'Method for the secure loading of a power system.' *Proc. IEE,* **115**.
20. J. H. Pickles (1975). 'Automatic load prediction by the spectral analysis method.' *Proc. Fifth Power Systems Computations Conf., Cambridge, 1975.*
21. E. D. Farmer and M. J. Potton (1968). 'Development of on-line load prediction techniques with results from the south-west region of the CEGB.' *Proc. IEE,* **115,** No. 10.
22. E. D. Farmer and M. J. Potton (1966). 'The prediction of load on a power system.' *Proc. Third IFAC Conf., London, 1966.*
23. F. Meslier (1978). 'New advances in short-term forecasting using the Box–Jenkins approach.' *IEEE Winter Power Meeting, PES, New York, 1978.*
24. M. J. H. Sterling and A. Bargiela. Chapter 12 of this volume.
25. E. D. Farmer (1980). 'The economics and dynamics of power system loading and regulation.' *IEEE Int. Conf. Power System Monitoring and Control, 1980. Conf. Publication No. 187.*
26. R. M. Dunnett (1983). 'Reserve allocation on a thermal power system with pumped-storage and gas-turbine capacities.' *IEEE Int. Conf. Power System Reliability, September 1983.*

27. A. M. Adatia, A. B. Baker, E. D. Farmer, W. D. Laing, and D. G. Robertson (1982). 'Interactive load forecasting and reserve allocation.' *Proc. C.I.G.R.E., September 1982.*
28. M. R. Irving and M. J. H. Sterling. Chapter 10 of this volume.

Comparative Models for Electrical Load Forecasting
Edited by D. W. Bunn and E. D. Farmer

CHAPTER 2

Review of Short-term Forecasting Methods in the Electric Power Industry

D. W. Bunn and E. D. Farmer

1 INTRODUCTION

The previous chapter has shown how the accuracy and timeliness of load forecasts can have a considerable effect on the operational economies of an electric power system. In this chapter we review current practices in the methods used to make such forecasts. In this study we are essentially concerned with those forecasts made routinely throughout the day for lead times varying from a few minutes to several hours which support the load dispatching and unit commitment decisions. There has recently been a most useful and extensive review of types of statistical modelling techniques used in this field of forecasting.[1] That review provides rather more detail of each type of statistical method (spectral, exponential smoothing, state-space, etc.) than we do here and reviews the literature according to the applications of each. Since most forecasting procedures developed by electric utilities appear to be a hybrid of several types of statistical procedure, we will develop a slightly different taxonomy of the methods relating more to the overall concept of the forecasting model and certain practical features of the information system.

Forecasts produced for longer lead times than we consider here are generally concerned with daily, weekly, seasonal, or annual peaks in the load[2–5] or the load duration curve.[6,7] For shorter lead times than are required for dispatching purposes, load flow analysis is mostly concerned with state estimation and validation for security checks on the system.[8,9]

2 DESIGN FEATURES

With the increasing development of automated load forecasting and dispatching algorithms in the power supply industry, certain design features of the short-term load predictor become important. These include:

(a) *Adaptiveness* At least some, if not all, of the parameters in the forecasting model will not remain stationary over time. Parameter changes should be tracked automatically in the overall model.

(b) *Recursiveness* As new data becomes available, possibly on a minute-by-minute basis, forecasts should be re-computed in a manner which reacts appropriately to new data but does not require the whole past history of data to re-estimate any parameters.

(c) *Computational economy* Even though power-supplying organizations are prepared to dedicate a great deal of computing resources to the operation of on-line load forecasting and dispatching, the short-term forecasting algorithm needs to be economical both with respect to execution time and core utilization. The back-up reserve requirement of having a second computer available in case of a breakdown incident often reinforces this need.

(d) *Robustness* There are three types of robustness to consider, namely:

(*i*) *Specification*

The model will often be operating upon a part of the time series for which it is locally misspecified and therefore sub-optimal. Reasonable forecasts should still be maintained in these circumstances.

(*ii*) *Measurement*

Measurement errors in the data are quite common and the model should be robust to these. This problem has usually motivated the development of a validity-checking routine for new data prior to processing.[8]

(*iii*) *Default*

The method should be robust to defaults in the input of new data either through system or operator failure. This is a particularly strong consideration for those procedures which require the periodic input of extra variables, such as weather data, by an operator in order to supplement the automatically logged load data. The forecasting procedure should have robust default procedures to deal with any less than perfect behaviour on the part of its 'exogenous agents'.

These design features are some of the basic considerations which have to be compromised in the formulation of short-term load forecasting procedures. No two power-supplying organizations are identical and the actual forecasting procedure which they implement will reflect the relative strength of importance of the above design features as well as some of their own operational idiosyncrasies. It is a reflection of the size of methodological choice in applied forecasting that even under the above design constraints, there is still a wide diversity of methods which have been suggested and tested for this short-term forecasting problem.

3 TAXONOMY

For the purposes of developing a loose taxonomy within these applied load forecasting methods, three important characteristics can be identified. Apart from other aspects, applied short-term load predictors differ according to whether they use a 'standard load curve', how they make use of weather variables, and whether the data is spot or integrated over a period of time.

The *standard load curve*, which is also often referred to as a 'nominal' or 'reference' load curve, has been a popular forecasting device. Usually, the standard load curve is produced on a daily basis, reflecting long-term average behaviour. Throughout a particular day, the residual deviations between the actual load and the standard load are computed and adaptively forecast. This concept of forecasting is very readily adopted by the control and operations management as their daily preoccupations with the shape of the load curve gives them a considerable (almost empathetic) insight into the determinants and nature of any changes in the daily pattern. With this resource of good subjective knowledge, on-line forecasting procedures would apparently have much to gain from the incorporation of interactive facilities, if not the explicit formulation of composite subjective and empirical evidence.[10] The standard load concept is one that readily admits the cooperation of the operational management and this is an immensely strong point in its favour. Several variations in the construction of the standard load curve have been evident; these will be described in the next section.

The effects of *weather variables* on the demand for electricity have been analysed in several studies.[11–15] In Great Britain,[16] the principal weather variables which influence short-term loads are temperature (at various lags), wind speed, and effective illumination (a function of cloud cover, visibility, and precipitation). In other countries, particularly those with a substantial air-conditioning component in the load, the additional factor of humidity (wet and dry bulb temperature) will be important. Whilst off-line weather-dependent forecasting models are essential for lead times of more than a few hours and are usually developed for the 'unit commitment' scheduling decisions, for the short-term context of this study their on-line implementation has been relatively limited. There are several reasons for this. In the first place, the on-line use of weather data for a real-time forecasting system requires either a reliable weather metering system of its own or the regular external input of weather variable forecasts from a meteorological service. Either of these procedures would be demanding in itself and would also require careful default routines to be developed in order to achieve the desired robustness in on-line operations. Secondly, it has been suggested[17] that univariate adaptive procedures will implicitly be tracking most weather-induced changes over the short-term and thus little extra predictive accuracy would be gained from constructing a more elaborate multivariate weather-dependent model. Pickles[18] has provided some empirical evidence to support this.

To a certain extent, the reasoning of load forecasters on the incorporation of weather data into short-term predictors is coloured by the direction from which they approach the problem. If the load dispatching is seen as a slightly shorter term operation than the unit-commitment exercise, the weather-dependent type of approach may be more likely to be preserved to the shorter term. Alternatively, if the automatic load dispatching is seen as a development from the minute-by-minute control and security validation function, then the adaptive univariate approach is likely to be stretched to the load dispatching lead times. The organization structure of these functions can clearly be influential in these trends. Certain hybrid formulations have also been advocated and these will be described in subsequent sections.

Whether the *load data* is spot or integrated over time (half-hourly or hourly) again reflects to a certain extent the possible origins of the load dispatching predictor in the unit-commitment or security validation functions. The forecasts for unit commitment at lead times of several hours ahead usually use integrated data (MWh) which is easier to meter than the extensive telemetering and load validation systems which are needed to provide minute-by-minute spot values of the load. Because of the smoothing effects in the integrated data, it is not so useful for precise load dispatching over the short-term. Thus, when reviewing load forecasting procedures, it is important to be aware of the nature of the input data.

In the following sections, some of the applied short-term load forecasting procedures will be presented under a broad categorization relating to the use of the standard load concept.

4 STANDARD LOAD FORMULATIONS

The essential concept of the standard load formulation involves modelling the actual load, $L(t, d)$, for time t of day d, in terms of a standard load, $S(t, d)$, and a residual, $R(t, d)$. Most commonly this decomposition has taken the additive form[17–27]

$$L(t, d) = S(t, d) + R(t, d). \tag{2.1}$$

Alternatively, a multiplicative formulation can be adopted:

$$L(t, d) = R(t, d)S(t, d) \tag{2.2}$$

as, for example, the current model in use by the Central Electricity Generating Board.[28]

For illustration, Figure 2.1 shows the actual and standard CEGB load curves for a randomly sampled day in 1980 (the standard load curve is the smoother of the two). The standard curve is produced once a day and although it is evidently of the right 'shape' (in this instance), it needs 're-scaling' over time (hence the importance of $R(t, d)$). Figure 2.2 shows the value of this re-scaling procedure through a series of sequenced 'snapshots' of the forecasts and actuals. Thus, at 16.30, the forecast from the on-line predictor can be seen for up to 3 hours ahead (again the smoother

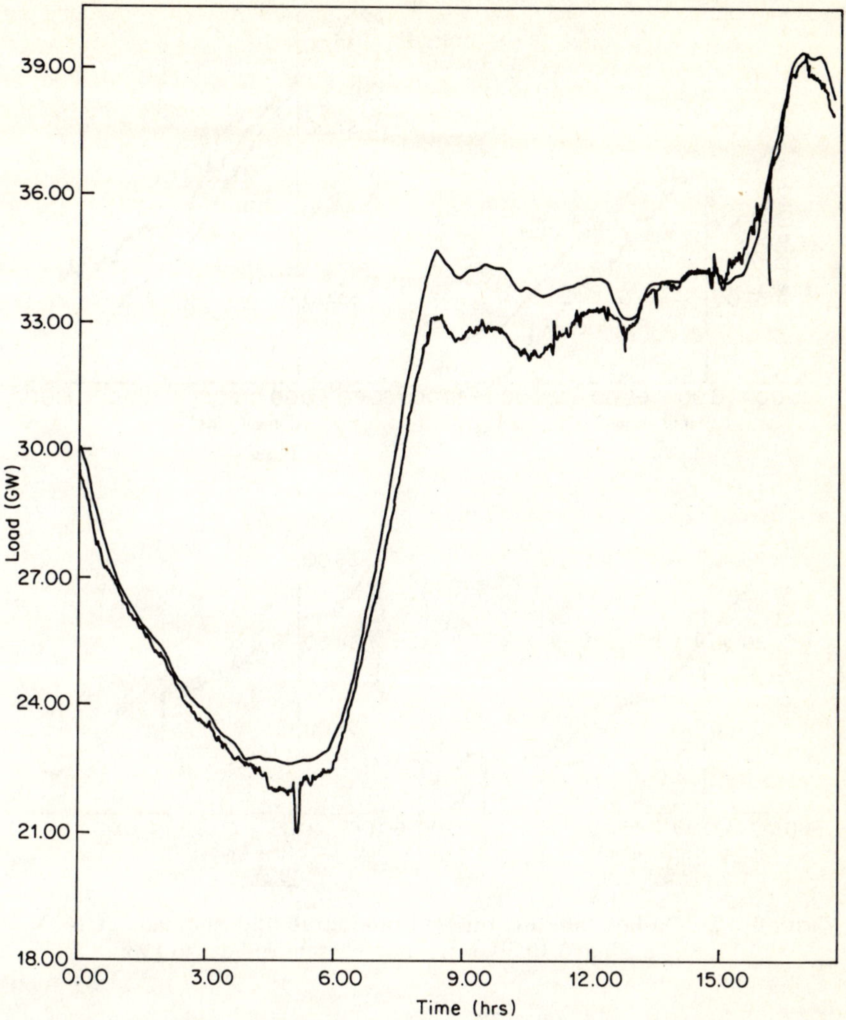

FIGURE 2.1 Actual and standard load curves for a randomly sampled Tuesday in 1980

curve) alongside the load profile which actually happened. Note that this on-line predictor is only designed and used for up to 30 min lead times; the longer lead times here are only presented for illustration. Then, at 16.30, the forecasts for up to 30 mins ahead were quite good but divergent afterwards. By 17.30, however, the forecasts had adapted correctly and this short-term adaptation can be seen in the subsequent plots for 18.30 and 19.30. Note that the $R(t, d)$ are fitted over the previous 60 min data, as

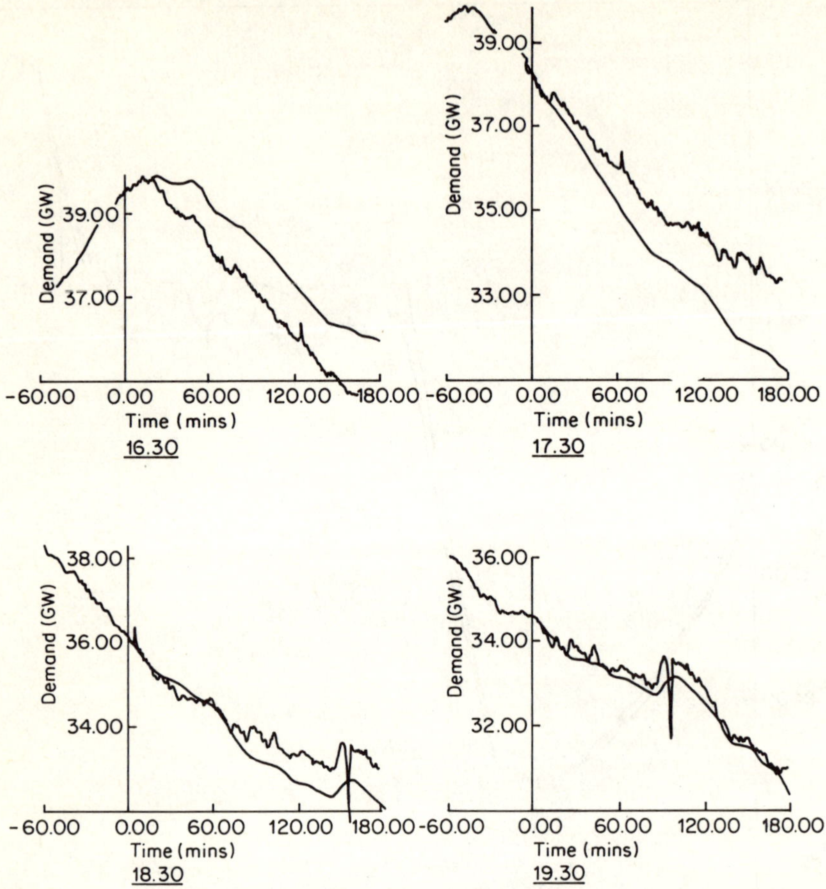

FIGURE 2.2 On-line forecasts for lead time up to 180 min made at 16.30, 17.30, 18.30, and 19.30 on a randomly sampled day in 1980

shown in the plots, whereas the $S(t, d)$ is based upon a long history of seasonal and daily patterns.

Within the basic standard load decomposition, a variety of methods have been proposed for estimating $S(t, d)$ and $R(t, d)$. The methods proposed for $S(t, d)$ will be discussed first. These can be divided into 'Historical' and 'Forecasted' estimators.

4.1 Historical $S(t, d)$

The historically-derived standard load curves essentially represent some form of a moving average representation. This $S(t, d)$ is usually broken down into at least two components:

$$S(t, d) = A(t, d) + D(t, d) \tag{2.3}$$

where $A(t, d)$ is the weekly or seasonal average and $D(t, d)$ a day-of-the-week effect. A simple moving average has been the most common method of estimating $A(t, d)$. Thus, if the past n weeks contains a total of N 'ordinary' days ($N \neq 7n$ because of the exclusion of 'special' days such as holidays), then

$$A(t, d) = \sum_{i=d-N}^{d-1} L(t, i)/N \tag{2.4}$$

if $A(t, d)$ is updated daily on the basis of an n-week moving average. Examples of this technique can be found in References 17–20 and 22. The parameter n has usually been fitted by simulations on past data.

A similar type of moving average has also been used to estimate the day-of-the-week effect, $D(t, d)$. Thus,

$$D(t, d) = \sum_{k=1}^{n} (L(t, d-7k) - A(t, d-7k))/n \tag{2.5}$$

if $D(t, d)$ is updated weekly on the basis of an n-week moving average. Examples of this approach can be found in References 17, 22, and 26. However, the use of an exponentially-weighted moving average appears to have been a more common estimator for $D(t, d)$. Thus,

$$D(t, d) = \alpha(L(t, d) - A(t, d)) + (1-\alpha)D(t, d-7) \tag{2.6}$$

where α is an empirically derived smoothing parameter. Various versions of this approach can be found in References 18–21 and 26.

Comparisons between these two approaches, simplified by equations (2.5) and (2.6), for estimating $D(t, d)$ have been equivocal. Lijesen *et al.*[26] indicated a preference for the moving average. However, the experience of the Central Electricity Generating Board seems to indicate a trend in favour of exponential smoothing. The early work,[17] using simple moving averages, was superseded[18] by an exponential smoothing approach. Furthermore the CEGB has now consolidated the categorization of $S(t, d)$ into $A(t, d)$ and $D(t, d)$ by using seven $S_i(t, d)$, representing the standard for each day of the week, with each $S_i(t, d)$ being updated separately by exponential smoothing.[28] Some work on relating $S_i(t, d)$ to $S_{i-1}(t, d-1)$ via a 'yesterday effect has not proven worthwhile. An advantage of using this consolidated exponential smoothing over the decomposition into additive $A(t, d)$ and $D(t, d)$ is that it does not assume the day-of-the-week effect to be a constant increment, independent of seasonality.

One disadvantage of both types of averaging is that the smoothed estimate inevitably lags the current time index. Thus, several research workers have attempted to incorporate a trend effect $T(t, d)$ into the decomposition of $S(t, d)$. Hence,

$$S(t, d) = A(t, d) + D(t, d) + T(t, d). \tag{2.7}$$

In Abu-Hussien *et al.*,[22] $T(t, d)$ is modelled as a second-order polynomial. Pickles[18] used an empirical function

$$T(t, d) = \tfrac{1}{2}nf(t, d)$$

where

$$f(t, d) = f_1(d) - F_1(t)/(\Delta_1^2(t) + (d - d_1(t))^2) + F_2(t)/(\Delta_2^2(t) + (d - d_2(t))^2) \tag{2.8}$$

such that $f_1(d)$ is a fourth-degree polynomial which describes the relatively smooth changes of $L(t, d)$ at middays and $F_i(t)$, $\Delta_i(t)$, and $d_i(t)$ model the additional effects due to changes in the times of sunrise and sunset during the spring and autumn. Other researchers have consolidated $T(t, d)$ in the estimation of $A(t, d)$. De Martino *et al.*[21] have achieved this through modelling $A(t, d)$ as a type of Holt–Winters[36] double exponential smoothing. Lijesen *et al.*[26] have modelled $A(t, d)$ as a function

$$A(t, d) = a(t) + b(t)n + c(t)\sin(2\pi n/52) + d(t)\cos(2\pi n/52) \tag{2.9}$$

with the parameters $a(t)$, $b(t)$, $c(t)$, and $d(t)$ estimated empirically. The adaptive, on-line estimation of these parameters was only partially resolved, however, in this work.

Further consolidation of $T(t, d)$, $A(t, d)$, and $D(t, d)$ has been undertaken in some publications[23–25] resulting in a functional approach to modelling $S(t, d)$. Galiana[23] models $S(t, d)$ as a finite Fourier series with a weekly period. Panuska[24,25] also uses a similar finite Fourier series but on a daily period. Hence, the Panuska[24,25] model would require separate day-of-the-week models to be developed. Takenawa *et al.*[41] have separate weekday and weekend models of periods 5 and 2 respectively.

The historically based standard load curves have been popular in practice for at least two reasons:

(i) They can generally lead to a satisfactory level of performance in meeting those on-line design features previously identified as desirable in short-term load predictors.
(ii) The production of average load curves is useful managerial information for other purposes apart from serving the load forecasting and dispatching algorithm.

However, the very fact that they are historical reduces their extrapolative power for lead times of an hour or more. Standard load curves which embody some forecasting on the basis of known relationships between weather variables and electricity demand would appear, therefore, attractive at this longer end of short-term load predictions.

4.2 Forecasted $S(t, d)$

A direct extension of the reasoning which formulated the decomposition of $S(t, d)$ into $A(t, d)$, $D(t, d)$, and $T(t, d)$ leads to the approach of decomposing the standard load into an additional weather-dependent component, $W(t, d)$. Thus, we would now have

$$S(t, d) = A(t, d) + D(t, D) + T(t, d) + W(t, d). \qquad (2.10)$$

Van Meeteren and van Son[19] and Abu-Hussein *et al.*[22] use this approach. The van Meeteren and van Son[19] paper describes a theoretical basis for incorporating $W(t, d)$ but only gives a very limited indication of its facility in terms of experimentally simulated temperature effects. Abu-Hussien[22] implemented a temperature-dependent component, $W(t, d)$ into an Egyptian context using the model

$$W(t, d) = \begin{cases} a_1 + b_1(T_{t-1} - 70) + c_1(T_{t-1} - 70)^2 & \text{if} \quad T > 70 \\ 0 & \text{if} \quad 70 \geq T \geq 60 \\ a_2 + b_2(60 - T_{t-1}) + c_2(60 - T_{t-1})^2 & \text{if} \quad T < 60 \end{cases} \qquad (2.11)$$

where T_t is the temperature (°F) at time t (in hours).

Krogh *et al.*[17] use a regression-based technique to 'normalize' daily peaks and troughs for weather variables and thus allow a univariate time-series model (ARIMA) to be used to the hourly normalized data. Irisarri *et al.*[48] developed an hourly-recursive Kalman filter, which includes an exogenous temperature forecast, to formulate their equivalent of an $S(t, d)$.

Ackerman[27] departs more radically from the historical model by constructing the short-term standard curve, $S(t, d)$, as a cubic interpolation of the hourly weather-dependent load predictor developed for unit-commitment decisions.

Whilst the use of forecasted standard curves provides a greater 'look ahead' capability[27] [sic] over the historically derived methods, this is gained at the expense of other desirable features. Because the standard loads depend upon the regular input of either weather measurements or forecasts, the procedure becomes less robust through the possibilities of breakdowns in this communications line. Bunn[29] has proposed a method which avoids this issue, but at the expense of greater computational costs. Essentially, a set of m standard loads $S_i(t, d)$ are constructed for different weather-induced conditions and probabilities, $P_i(t)$, are defined such that the marginal $S(t, d)$ is evaluated as an expectation

$$S(t, d) = \sum_{i=1}^{m} P_i(t) S_i(t, d). \tag{2.12}$$

The $P_i(t)$ are revised in a Bayesian manner as each new $L(t, d)$ value becomes available throughout the day.

Having reviewed various approaches for the estimation of the standard load curve, it is now appropriate to consider the other half of this formulation, namely the adaptive estimation of the residuals, $R(t)$. Again, it is useful to distinguish between univariate methods which use only the past load data and multivariate procedures which make use of weather variables or other exogenous inputs (e.g. subjective forecasts).

4.3 Univariate methods

(i) Autoregressive

A common approach has been the autoregressive model AR(k):

$$R(t) = \sum_{i=1}^{k} \alpha_i R(t-i) + \varepsilon(t) \tag{2.13}$$

where $\varepsilon(t)$ is white noise residual. On hourly data, Galiana[23] has used AR(1) and AR(2) for two different residual series. Gupta[20] has used AR(24) on hourly data but has been criticized in the discussion of that paper for high parametrization.

It is interesting to observe that authors taking an ARIMA approach to model identification have also eventually adopted AR(1) procedures, e.g. Ackerman[27] on five-minute and van Meeteren *et al.*[19] on hourly spot data.

(ii) Exponential smoothing

Simplicity, parsimony, recursiveness and computational economy are attractive features of exponential smoothing which have encouraged its implementation. Single exponential smoothing was used in Farmer and Potton.[17] A form of double smoothing is currently in use by the CEGB,[28] the motivation of which is to pick up short-term trends more effectively. Two single exponentially smoothed values for $R(t)$ are obtained from different smoothing constants α_1 and α_2, namely $R_1(t)$ and $R_2(t)$. For lead time L, the estimated residual $\hat{R}(t+L)$ is given as

$$\hat{R}(t+L) = R_1(t) + w(t, L) f(\alpha_1, \alpha_2, L)(R_1(t) - R_2(t)) \tag{2.14}$$

where $w(t, L)$ is a function of the lead time, L, and the rate of change of the underlying standard curve, $S(t, d)$; and $f(\alpha_1, \alpha_2, L)$ is a function of the lead time and the two different smoothing constants. The effect of $w(t, L)$ is to reduce the extrapolation of trends over longer lead times. The parameters of this procedure are thus adaptive with respect to the changes in the standard load, but not with respect to the previous forecasting errors.

(iii) Spectral

Some investigations have been undertaken using an expansion of $R(t)$ on terms of the eigenvectors of its autocovariance matrix.[17,18] Thus,

$$R(t) = \sum_{i=1}^{k} a_i \lambda_i^{1/2} Q_i(t) + \varepsilon(t) \tag{2.15}$$

where $Q_i(t)$ and λ_i are the associated ith eigenvector and eigenvalue of the autocovariance matrix, $\boldsymbol{C}(t)$ where

$$\boldsymbol{C}(t) = \mathbf{r}(t)\mathbf{r}^{\mathrm{T}}(t) \tag{2.16}$$

with $\mathbf{r}(t)$ the $k \times 1$ vector of the previous k observations to time t. A recursive estimation procedure is available for $R(t)$.[17,18] Some experience of testing this method in Britain and Japan has been reported,[17] notable features being the sensitivity of the method to the estimation of its parameters and its lack of robustness relative to exponential smoothing under conditions of high measurement noise.

4.4 Multivariate methods

Modelling the residuals, $R(t)$, in a multivariate form has not been common in practice. This is mainly because the underlying rationale of the standard load curve is that the residuals should be adapting quickly to short-term variations in the load. Thus, a multivariate model generally requires a large amount of regular exogenous data input, such as weather measurements. To do this reliably would involve incorporating on-line weather-variable measurements, which may be prohibitively expensive.

However, Panuska,[24,25] extending some previous work of Galiana,[23] has incorporated a lagged temperature-variable into an ARIMA type of predictor for hourly residuals. Lijesen and Rosing[26] incorporate a weather-sensitive component into their residuals by means of allocating a daily energy requirement (computed off-line) via a 24-element vector of constants to each hour. Takenawa *et al.*[41] used hourly temperature measurements to update a systematic temperature component which was of a nonlinear autoregressive form and this provided the adaptive feature in their model.

Within this class of multivariate methods, it is possible also to include procedures which rely upon the periodic supplementation of an on-line univariate procedure with off-line forecasts. The method of Gupta and Yamada[20] seeks to combine the daily peak forecast, produced off-line by a multivariate procedure, with an on-line univariate predictor. The combination is optimal-seeking, although it does not take into account the correlation between the off-line and on-line forecasts. However, this correlation can be quite significant (cf. Pickles[18]). Farmer *et al.*[42] have also developed a procedure for extending the 30 min lead time of the CEGB on-line predictor to a couple of hours through using the off-line 'unit-commitment' forecasts as 'targets' in the extrapolation of the residual function (recall the divergence of the on-line forecasts for longer lead times in Figure 2.2).

5 UNSTANDARDIZED LOAD METHODS

Insomuch as nearly all the models which have been developed for short-term load forecasting are linear, the distinction between standard and unstandardized load models may appear to be one only of interpretation. The linear model will contain a subset of components relating to periodicities and trends of duration greater than a day, which could be conceptualized as being a functional representation of a standard load curve. The remaining subset modelling daily variations is analogous to the adaptive forecasting of residuals in the standard load formulation. However, an important point of distinction is in the motivation behind the model. The standard load approach explicitly seeks to provide a daily series of residuals which will be adaptively tracked and forecast, whereas the unstandardized load approach seeks to identify an overall model to give white noise residuals. A further point of distinction relates to the degree of recursiveness in the sequential estimation of the model. In the standard load approach, the ideal is to update the standard load curve once a day, and the residual parameters every minute or so. The full recursive operation of the unstandardized proceeding would require the whole model to be updated at every minute or so. Finally, an important point of distinction for the practical implementation of a short-term load forecasting procedure is whether it is seen to make explicit use of a standard load curve. It appears to be the case that the literature contains more examples of the practical success of explicitly standard load approaches than of unstandardized methods. This may reflect a greater managerial support for a more intuitively appealing formulation.[40]

Applications of unstandardized load procedures can be considered according to the type of time-series methodology formulated.

(a) *Autoregressive*

Srinivasan and Pronovost[34] have applied an unusual form of autoregressive model for hourly data:

$$\hat{L}(t, d) = \sum_{k=1}^{4} \alpha_k \hat{L}_k(t, d) \qquad (2.17)$$

where

$\hat{L}_1(t, d)$ is a forecast of $L(t, d)$ based upon first-order autoregression with a lag of one hour

$\hat{L}_2(t, d)$ is a forecast of $L(t, d)$ based upon first-order autoregression with a lag of one day

$\hat{L}_3(t, d)$ is a forecast of $L(t, d)$ based upon first-order autoregression with a lag of one week

$\hat{L}_4(t, d)$ is a forecast of $L(t, d)$ based upon first-order autoregression with a lag of one year

and the α_k are linear weights seeking an optimal combination of these four separate forecasts (although covariances were not taken into account). The explicit nature of this formulation in terms of its being a combination of four separate autoregressive models may have been motivated by managerial appeal.

(b) ARIMA

Keyhani and Rad,[33] Abu-el-Magd and Sinha,[43] Meslier,[44] Vemuri *et al.*,[45] and Hagan and Klein[37] have applied the ARIMA time-series approach of Box and Jenkins.[38] Vemuri *et al.*[45] have presented a sequential identification method whilst Keyhani and Rad[33] have additionally incorporated temperature variables into the formulation.

(c) General exponential smoothing

Christiaanse[30] has adapted the generalized exponential smoothing methods of Brown[39] forecasting hourly integrated loads:

$$L(t) = \mathbf{a}^{\mathrm{T}}\mathbf{f}(t) + \varepsilon(t) \tag{2.18}$$

where $\mathbf{f}(t)$ is a vector of 'fitting functions' and $\mathbf{a}$ is a vector of exponentially smoothed weights. The fitting functions were expansions of a Fourier series based upon a period of one week.

(d) Kalman filtering

Sharma and Mahalanabis[31] have found that Kalman filtering of the $\mathbf{a}$ vector in Christiaanse's formulation,[30] equation (2.18), gave improved forecasts. Abu-el-Magd and Sinha[43] have compared a Kalman filter with an ARIMA model, concluding that although their performances were similar, the Kalman approach may be preferable because of its greater flexibility of formulation. Toyoda *et al.*[32] have also discussed various Kalman filtering procedures in the short-term load forecasting context.

(e) Pattern recognition

Matthewman and Nicholson[35] have presented a pattern-recognition approach, although their results were not encouraging.

6 FORECASTING SPECIAL EFFECTS

Comparative research into forecasting methods has commonly focused upon the properties of optimality, robustness, and computability. Very little published consideration has been given to the implementation of forecasting procedures and, in particular, to the issues of 'tailoring' a standard technique to the peculiarites of a specific practical application. This section summarizes some of the 'special effects' which have had to be taken into account when implementing a short-term, on-line forecasting procedure for electricity loads. In most cases, special subroutines have had to be developed for each of these effects, the construction of which may have had more influence upon the overall forecasting performance than the choice of the basic, underlying predictor.

(*a*) *Weather factors* Such variables as temperature, windspeed, visibility, illumination level, and humidity can affect short-term variations in the demand for electricity. Many of these are modelled systematically as discussed previously, but sometimes special ad hoc routines have to be developed to deal with such events as the onset of an off-shore breeze or a sequence of hot days.[41]

(*b*) *Television audience behaviour* Popular television programmes can cause very significant effects on the evening demand pattern. A study of this problem in the UK has recently been undertaken.[46]

(*c*) *Holidays* National holidays cause discontinuities in the automatic operation of a load predictor. Christiaanse[30] suggested the suppression of error feedback on holidays to avoid upsetting ordinary parameter estimation. Other procedures for this problem are described elsewhere.[28,30,34,41] Even school holidays can cause certain changes to the load profile (particularly midmorning).

(*d*) *Onset of darkness* Predicting the onset time and extent of the evening lighting load has had an important effect upon the performance of the CEGB's on-line predictor[28] and the San Diego model.[41]

(*e*) *Daylight-saving time* The clock changeovers from summertime to wineterime have lagged effects which need to be modelled separately.[17]

(*f*) *Load management* With the increasing use of load management by electric utilities, forecasts of the 'unmanaged' load may have to be maintained as well as those monitoring the 'managed' load.

(*g*) *Data validation* Checking the validity of telemetered data is an important pre-processing function to avoid data bank contamination and poor forecasting.[28,30]

(*h*) *Industrial load monitoring* The demands of large industries should be separately monitored for shutdowns, strikes, or unusual patterns of working.[28]

(*i*) *Interactive features* Managers and control engineers should have easy interaction with the forecasting system to facilitate the incorporation of useful peripheral information, extra forecasts, etc. Consideration should, therefore, be given to the quality of output from the model, e.g. graphical displays, confidence intervals, etc. The CEGB in the UK is currently extending its interactive facilities.[42]

7 CONCLUSION

It is evident that no single most preferable forecasting method has emerged for the short-term load forecasting problem. Although all electricity-supplying organizations are confronted with very similar situations, the literature on the subject reveals a very wide spectrum of approaches.

One of the reasons for this is that a considerable amount of 'tailoring' has to be done to the basic predictor in rather ad hoc ways to deal with a variety of special irregular effects on the time series.

Furthermore, the current state of load dispatching methodology in practice is such that the on-line forecasting procedures generally serve only to *advise* control engineers. The advance of automation has not yet reached the stage where the whole forecasting and dispatching system is devoid of manual intervention. Thus, the most effective procedures are those which are intuitively attractive in their logic and readily interactive with the judgement of control engineers. Given the variety of styles and sophistications of operators, it is perhaps natural to expect a wide variety of favoured procedures to emerge. It has been the purpose of this volume to solicit contributions that would help to provide more insight into the practical and algorithmic considerations which have led to the favouring of particular approaches.

REFERENCES

1. M. A. Abu el-Magd and N. K. Sinha (1982). 'Short-term load demand modeling and forecasting: A review.' *IEEE Trans.*, **SMC-12,** No. 3, 370–382.
2. N. D. Uri (1979). 'An integrated model for forecasting system load.' *Applied Mathematical Modelling*, **3,** No. 1, 37–40.
3. P. C. Gupta (1971). 'A stochastic approach to peak power-demand forecasting in electric utility systems.' *IEEE Trans.*, **PAS-90,** No. 2, 825–832.
4. K. N. Stanton (1970). 'Medium-range, weekly and seasonal peak demand forecasting by probability methods.' *IEEE Summer Power Meeting, Los Angeles, 1970.* pp. 1183–1189.

5. K. N. Stanton and P. C. Gupta (1970). 'Forecasting annual or seasonal peak demand in electric utility systems.' *IEEE Trans.*, **PAS-89,** No. 5/6, 951–959.
6. N. D. Uri and J. S. Maybee (1980). 'Long term forecasts of optimal approximations to the load duration curve.' *J. Operational Research Society*, **31,** No. 4, 343–348.
7. J. S. Maybee and N. D. Uri (1980). 'Long term forecasting of smooth approximations to the load duration curve.' *Applied Mathematical Modelling*, **4,** No. 2, 130–132.
8. E. D. Farmer, P. Jervis, W. D. Laing, and J. N. Prewett (1974). 'Power system state estimation for load dispatching.' *Conference Internationale des Grands Réseaux Electriques, Paris, 1974.*
9. J. Zaborsky, T. Y. Chiang, and A. K. Subramanian (1977). 'Modelling and estimation of the load demand of an area.' *IEEE Summer Power Meeting, Mexico City, 1977.*
10. W. G. Michaelson and R. B. Comerford (1977). 'Forecasting and computers—the multi-model approach.' *IEEE, PICA*, 1977. pp. 247–251.
11. M. Davies. (1958). *The relationship between weather and electricity demand*, Monograph 3145, IEE, London.
12. G. T. Heinemann, D. A. Nordman, and E. C. Plant (1966). 'The relationship between summer weather and summer loads.' *IEEE Trans.*, **PAS-85,** No. 11 1144–1154.
13. C. E. Asbury (1975). 'Weather load model for electric demand and energy forecasting.' *IEEE Trans.*, **PAS-94,** No. 4, 1111–1116.
14. S. L. Corpening, N. D. Reppen, and R. J. Ringlee (1973). 'Experience with weather sensitive load models for short and long-term forecasting.' *IEEE Winter Power Meeting, New York, 1973*, pp. 1966–1972.
15. J. J. Saaks, G. W. Cunningham, and K. W. Priest (1972). 'Practical application of weather sensitive load forecasting to system power' IEEE Summer Power Meeting, San Françisco, 1972. pp. 971–977.
16. A. B. Baker. Chapter 5 of this volume.
17. E. D. Farmer and M. J. Potton (1968). 'Development of online load-prediction techniques with results from the south-west region of the C.E.G.B.' *Proc. IEE*, **115,** No. 10, 1549–1558.
18. J. H. Pickles (1974) *Automatic load prediction by the spectral analysis method*, RD/L/N115/74, Central Electricity Research Laboratories, Leatherhead.
19. H. P. van Meeteren and P. J. M. van Son (1979). 'Short-term load prediction with a combination of different models.' *IEEE PICA, 1979*, pp. 192–197.
20. P. C. Gupta and K. Yamada (1972). 'Adaptive short-term forecasting of hourly loads using weather information.' *IEEE Winter Power Meeting, New York, 1972.* pp. 2085–2094.
21. B. De Martino, G. Fusco, E. Mariani, R. Randino, and P. Ricci. 'A medium and short-term forecasting model for the electrical industry.' *IEEE PICA, 1979.* pp. 187–191.
22. M. S. Abu-Hussien, M. S. Kandil, M. A. Tantuary, and S. A. Farghal (1979). 'An accurate model for short-term load forecasting.' *IEEE Winter Power Meeting, 1979.*
23. F. D. Galiana, E. Handschin, and A. Fiechter (1974). 'Identification of stochastic electric load models from physical data.' *IEEE Trans.*, **AC-19,** No. 6, 887–893.
24. V. Panuska (1977). 'Short-term forecasting of electric power system load from a weather-dependent model.' *IFAC Symposium, Melbourne, 1977*. pp. 414–418.

25. V. Panuska and J. P. Koutchonk (1978). 'Electrical power system load modelling by a two-stage stochastic approximation procedure.' *Proc. 6th Triennial World Congress of IFAC, Boston, 1975.*
26. D. P. Lijesen and J. Rosing (1970). 'Adaptive forecasting of hourly loads based upon load measurements and weather information.' *IEEE Winter Power Meeting, New York*, 1970. pp. 1757–1767.
27. G. B. Ackerman. Chapter 3 of this volume.
28. A. B. Baker, E. D. Farmer, W. D. Laing, and A. D. N. March (1978). *The online demand validation and prediction facility at the National Control Centre*, OD(S)/R38/78, Central Electricity Generating Board.
29. D. W. Bunn (1980). 'Experimental study of a Bayesian method for daily electricity load forecasting.' *Applied Mathematical Modelling*, **4,** No. 2, 113–116.
30. W. R. Christiaanse (1971). 'Short-term load forecasting using general exponential smoothing.' *IEEE Trans.*, **PAS-90,** No. 2, 900–910.
31. K. C. S. Sharma and A. K. Mahalanabis (1974). 'Recursive short-term load forecasting algorithm.' *Proc. IEE*, **121,** No. 1, 59–62.
32. J. Toyoda, M. Chen, and Y. Inoue (1970). 'An application of state estimation to short-term load forecasting.' *IEEE Trans.*, **PAS-89,** No. 7, 1678–1688.
33. A. Keyhani and T. Eliassi Rad (1977). 'A simulation study for recursive prediction of hourly load demands.' *IEEE PICA, 1977*. pp. 228–236.
34. K. Srinivasan and R. Pronovost (1975). 'Short-term load forecasting using multiple correlation models.' *IEEE Trans.*, **PAS-94,** No. 5, 1854–1858.
35. P. D. Matthewman and H. Nicholson (1968). 'Techniques for load prediction in the electricity supply industry.' *Proc. IEE*, **115,** No. 10, 1451–1460.
36. S. Makridakis and S. Wheelwright (1978). *Interactive forecasting*, Holden-Day, San Francisco.
37. M. Hagan and R. Klein (1977). 'Identification techniques of Box and Jenkins applied to the problem of short-term load forecasts.' *IEEE Summer Power Meeting, Mexico City, 1977.*
38. G. E. P. Box and G. M. Jenkins (1970). *Time-series analysis: Forecasting and control.* Holden-Day, San Francisco.
39. R. G. Brown (1965). *Smoothing, Forecasting and Prediction of Discrete Time Series.* Prentice-Hall, New Jersey.
39. R. G. Brown (1965). *Smoothing, Forecasting and Prediction of Discrete Time Series.* Prentice-Hall, New Jersey.
40. P. C. Gupta. Comments on the paper by Christiaanse (Reference 30).
41. T. Takenawa, A. M. Schneider and D. Schiffman. Chapter 7 of this volume.
42. E. D. Farmer, W. D. Laing, A. M. Adatia, A. B. Baker,and D. W. Bunn (1981). 'The development and implementation of an interactive demand validation and prediction facility.' *Seventh PSCC, Lausanne, 1981.*
43. M. A. Abu-El-Magd and N. Sinha (1981). 'Two new algorithms for on-line modelling and forecasting of the load demand of a multinode power system.' *IEEE–PES Winter Meeting, Atlanta, 1981.*
44. F. Meslier (1978). 'A new advances in Short-term load forecasting using the Box and Jenkins approach.' *IEEE–PS Winter Meeting, New York, 1978.*
45. S. Vemuri, W. C. Huang, and D. J. Nelson (1981). 'On-line algorithms for forecasting hourly loads of an electric utility.' *IEEE–PES Winter Meeting, Atlanta, 1981.*
46. D. W. Bunn and J. P. Seigal (1983). 'Television effects on the demand for electricity.' *Energy Economics*, **5,** No. 1, 31–36.

47. B. Krogh, E. S. de Llinas, and B. Lesser (1982). 'Design and implementation of an online load forecasting algorithm.' *IEEE Trans.*, **PAS-101,** No. 9, 3284–3289.
48. G. B. Irisarri, S. E. Widergren, and P. D. Yehsakul (1982). 'Online load forecasting for energy control center application.' *IEEE Trans.*, **PAS-101,** No. 1, 71–78.

PART TWO: COMPARATIVE ELECTRIC LOAD FORECASTING MODELS

Comparative Models for Electrical Load Forecasting
Edited by D. W. Bunn and E. D. Farmer

CHAPTER 3

Short-term Load Prediction for Electric-utility Control of Generating Units

Gary Ackerman

1 INTRODUCTION

Electric utilities predict the amount of energy and power that must be delivered to their customers for a variety of time horizons. Long-term forecasts covering twenty years or more are used by planning engineers to determine the type and size of generating plants that minimize both fixed and variable costs. Mid-range forecasts over one to three years are used by utilities to purchase fuel quantities and to assess revenue impacts due to changes in electricity tariffs. Finally, short-term forecasts, either minute-by-minute or hour-by-hour, are used by electric utility operators to control (i.e. schedule and dispatch) generating units. The objective of the operators is to minimize variable costs without jeopardizing the electric system to power failures.

Until recently, utility system operators or dispatchers relied on static prediction methods or judgment based on current load. There was no look-ahead capability over future time horizons (1–3 hours) to help determine the economic allocation of generation to the load.

The US Department of Energy sponsored a demonstration project to develop an advanced software package for automatic generation control (AGC). Systems Control, Inc., Wisconsin Electric Power Co. (WEPCO), and Control Data Corporation jointly conducted the different tasks of this effort. One of the tasks, undertaken by Systems Control, Inc., was to design and implement a short-term dynamic load predictor that could be empirically tested using WEPCO data. The results of that work for the hourly load predictor are recounted herein, with particular emphasis on the stages of model development from the initial hypothesized model to its final version.

2 SPECIFICATION AND EMPIRICAL DETERMINATION OF THE LOAD PREDICTORS

2.1 General model for time behaviour of load

A short-term electrical load predictor follows from the structure of the model used to represent the actual load evolution over time, and, in general, is given by a nonlinear function

$$y_t = f(y^t, \chi^t, t, \varepsilon_t) \tag{3.1}$$

where

y_t = actual system load in megawatts (expressed as integrated hourly load in the case of the hourly predictor)
$y^t = \{y_\tau; -\infty < \tau \leqslant t-1\}$; i.e. all past observed actual system loads
$\chi^t = \{\chi_\tau; -\infty < \tau \leqslant t-1\}$; i.e. all past observed exogenous variables
t = time index
ε_t = additive random disturbances representing all unobserved effects on the system load.

Attempting to develop a precise representation for $f(\cdot)$ can be very costly and although a very accurate predictor would result, it is an open question whether the improvement in accuracy over a more simple representation would justify the extra modelling effort. It may be possible to obtain sufficient accuracy with a simple representation. Therefore, it is advisable to begin with a simplified form for $f(\cdot)$ and elaborate upon it only as required in order to meet performance requirements. Such an approach leads one to the trivial representation:

$$y_t = \chi_t + e_t \tag{3.2}$$

where χ_t is now a scalar variable representing the value of the load at t as predicted by a regression of the load on all observed exogenous factors. The new additive random (unobserved) error, e_t, now represents both the truly random disturbances and the modelling errors inherent in replacing $f(\cdot)$ with χ_t. For the purposes of the hourly predictor and the chosen test site of this project, Wisconsin Electric Power Company (WEPCO), χ_t is taken as the WEPCO hourly integrated load forecast and is computed as an aggregate of all causal effects, including weather. More specifically, the WEPCO hourly load forecasting algorithm forecasts a base component and a weather-sensitive component of the load. The base component is updated by exponentially smoothing the actual loads for the same hours of the week. The weather-sensitive component is produced by a stochastic model whose inputs are temperature and dewpoint forecast data from two weather stations.

The optimal predictor for y_t can be obtained by formally taking conditional expectations across equation (3.2):

$$Y_{t+k|t-1} = \chi_{t+k} + e_{t+k|t-1} \tag{3.3}$$

where $y_{t+k|t-1}$ denotes $E\{y_{t+k} \mid I^{t-1}\}$ and I^{t-1} symbolizes all the information available at time $t-1$; likewise for $e_{t+k|t-1}$. The conditional expectation on χ_{t+k} reduces to χ_{t+k} itself since it is assumed that this value is always known into the future. From equation (3.2) it is seen that the structure imposed by equation (3.1) reduces the development of the optimal predictor to the development of an optimal predictor for e_t.

All the known structure, representing all *a priori* information concerning observed cause-and-effect relationships, is captured in the χ_t variables. All the unknown structure, representing all the *additional* information which may be extracted from empirical data, is contained in the e_t variable. If systematic (serial correlation, for example) variation exists in these error terms then it can be employed to devise additional structure which will yield a more accurate forecast. The examination of e_t is a problem in statistical time series analysis and is discussed below.

2.2 Empirical structure determination

The analysis of the error term in equation (3.2), and the estimation of a model for prediction of the error terms, constitute a problem in statistical time series analysis. The details of the methodology employed are given in Reference 1 and will not be presented here. Only a summary of the relevant steps comprising the methodology is given to facilitate an understanding of the rationale behind the predictors selected for implementation.

In short, the Box–Jenkins methodology is an iterative procedure by which a model is constructed. The process proceeds from the most simple structure, with the least number of parameters, to as complex a structure as is required to obtain an 'adequate' model—'adequate' in the sense of yielding white residuals. This process of building increasingly complex models embodies its own philosophy of parsimony: 'include only as many parameters as you really need.' A schematic of the procedure is given in Figure 3.1. The first step is an identification of structure and employs sample autocorrelation patterns. After a structure has been chosen the next step involves an estimation of the coefficients inherent in the structure description. Next the optimal parameter estimates are inserted into the model to generate its estimated residuals. These are then subjected to diagnostic procedures to determine if they are indeed 'white'. If not, their sample correlogram is used to hypothesize a new structure and the cycle is begun anew. If the model satisfies all diagnostic tests it may then be implemented for on-line testing. The benefits of such a methodology are many, but primarily one will always be assured of a model which has the *fewest* possible parameters while still explaining all the *systematic* variation in the random errors.

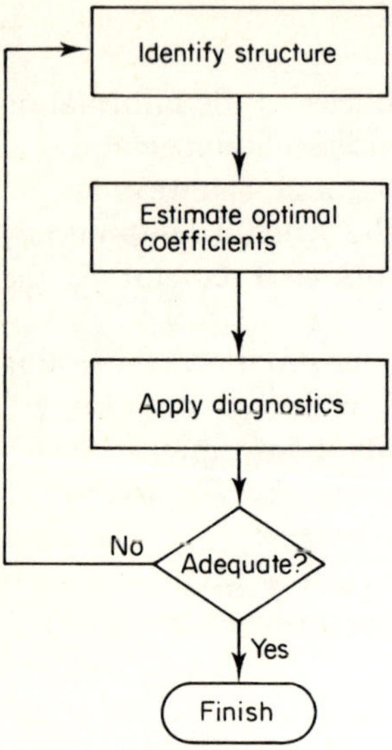

FIGURE 3.1 The Box–Jenkins methodology

The particular form for e_t assumed by the Box–Jenkins methodology is the rational form:

$$e_t = \frac{c(L)}{d(L)} \eta_t \tag{3.4}$$

where

$$c(L) = 1 + c_1L + \ldots + c_qL^q \tag{3.5}$$

$$d(L) = 1 + d_1L + \ldots + d_pL^p \tag{3.6}$$

where L is a shift (or lag) operator, i.e. $L^k\chi_t = \chi_{t-k}$, and η_t is a white noise process with the same normal distribution for every value of t. Thus, by structure identification we mean specification of the integers p and q. In the econometric terminology, equation (3.4) represents an auto-regressive/moving-average ARMA(p, q) model. By examining the auto-correlation function for e_t, it is possible to gain information regarding the values of p and q. This is the first step in the procedure.

Estimation of a set of coefficients $\{c_i, d_i\}$ is effected with the aid of maximum likelihood estimation algorithm.[2]

3 THE HOURLY LOAD PREDICTOR: EVOLUTION OF THE TIME SERIES MODEL

Three different specifications of an hourly load predictor model were examined. The specifications were as follows:

- *Specification I.* Based on preliminary hourly load data for January and

March, a twenty-fourth order differencing of the raw data was modelled as a first-order autoregressive (AR) process.

- *Specification II.* The data were consecutively differenced, i.e. $e_t - e_{t-1}$, and modelled as a twenty-fourth order autoregressive process combined with a first-order moving average (MA) of the residual error terms.
- *Specification III.* A different set of raw data was constructed to eliminate the input of the load dispatcher in the hourly forecast. A simple autoregressive (AR) model was identified as being robust across different data sets.

The different specifications were due to subsequent attempts to improve the hourly load predictor from the initial specification. Note that the third specification required a change in the raw data series since it was felt that the prior specifications exhibited residual error terms that were nonstationary. That is to say, estimation of the models exhibited wide variations in the parameters for different independent data sets.

The different raw data sets used in Specifications I and II versus that of III would, some might argue, make the comparison of parameter estimates meaningless. The estimates are conditional upon the model structure and the data. The purpose, however, in exposing the model development process for the hourly predictor is to demonstrate that specifying a model requires progressive steps, where at each step the modeller learns something new which is then incorporated in the succeeding variant. In the case of WEPCO's hourly loads, it was discovered upon reviewing the disappointing results of Specifications I and II that the raw data had been 'polluted' by one or more of the load dispatchers. The predicted hourly load from the model shown in equation (3.3) above, $\hat{y}_{t+k|t-1}$, was differenced from the observed value for y_{t+k} to derive $\hat{e}_{t+k}$. However, at various time periods in the data sets, the WEPCO dispatchers changed the predicted values, $\hat{y}_{t+k}$, to reflect their judgment regarding future load conditions. The problem, of course, is that the sporadic injection of human judgment, although improving the WEPCO hourly forecast, introduced a new component to the time series that was neither deterministic nor random. Hence, rather than trying to model the marginal accuracy of the load dispatcher (and we were never certain that the same load dispatcher was making the adjustments in every case), the original predicted value $\hat{y}_{t+k|t-1}$ was used for all hours.

The maximum likelihood results for the three specifications can be found in Table 3.1. All six data sets are consecutive weekdays. The June and August sets are accompanied by dates since there were two samples taken from each month, respectively.

What appeared to be an attractive specification for January and March data with Specification I failed to satisfy two requirements. First, application of the twenty-fourth order differencing to later data sets from June and

TABLE 3.1 Full information maximum likelihood estimation of time series specifications on WEPCO hourly loads

Specification I $e_t - e_{t-24} = \frac{1}{1+d_1L}\eta_t$

Data set	Number of observations	d_1	$\hat{\sigma}d_1$	$\hat{\sigma}_\eta$	R^2
January	96	−0.76	0.06	26.51	48.34
March	72	−0.66	0.09	30.76	43.16
June 5–9	120	−0.32	0.09	72.52	21.21
June 12–16	120	−0.40	0.09	75.47	14.67
August 7–10	96	−0.87	0.05	53.80	75.09
August 14–17	95	−0.85	0.05	42.73	72.65

Specification II $e_t - e_{t-1} = d_1L^{24}(e_t - e_{t-1}) + (1 + c_1L^1)\eta_t$

Data set	d_1	$\hat{\sigma}d_1$	c_1	$\hat{\sigma}c_1$	$\hat{\sigma}_\eta$	R^2
January	0.54	0.10	−0.28	0.13	24.01	31.92
March	0.39	0.12	−0.35	0.16	31.16	15.07
June 5–9	0.16	0.10	−0.79	0.07	59.12	21.20
June 12–16	0.04	0.10	−0.73	0.09	58.45	24.57
August 7–10	0.23	0.12	−0.29	0.10	49.02	0.0
August 14–17	0.66	0.10	0.07	0.10	48.70	29.98

Specification III $e_t = \frac{1}{1+d_1L}\eta^t$

Data set	d_1	$\hat{\sigma}d_1$	$\hat{\sigma}_\eta$	R^2
January	−0.78	0.05	12.49	89.91
June 5–9	−0.78	0.06	40.20	60.75
June 12–16	−0.82	0.05	41.79	66.42
August 7–10	−0.86	0.05	41.33	73.47
August 14–17	−0.76	0.07	63.37	55.89

Notes: d_1 is an autoregressive parameter; c_1 is a moving average parameter; $\hat{\sigma}$ is the estimated standard error; L^n is a polynomial lag operator of order n such that

$$[L^n e_t = e_{t-n}\cdot]$$

R^2 is the percentage variation in the dependent variable explained by the model.

August 1978 weekdays exhibited nonstationary properties in the sample autocorrelation functions. This indicated either that twenty-fourth order differencing might be appropriate for some but not all data sets, or that an alternative specification was required. Second, estimation of the AR(1) model on the twenty-fourth order differences exhibited parameter estimates for the autoregressive coefficient which were significantly different from the estimates made on January and March data (see left-most portion of Table 3.1).

Specification II exhibited widely varying parameter estimates for the autoregressive term. The range of values for the parameter labelled c_1, the moving average term, were from −0.79 to 0.07. This implied that for some data sets the model might be ARMA(24, 1) (e.g. January), AR(24) (e.g. August 14–17), or MA(1) (e.g. June 12–16).

Specification III was more robust across different data sets than the other two specifications. It should be noted that the sample autocorrelations of the noise process, η_t, indicated that greater complexity could be added to the AR(1) model. However, experimentation with more complex models never reduced the residual standard error by more than 5%, thus the AR(1) model was selected to represent the hourly load predictor error process. In order to confirm the conclusion presented above, one additional data set for August 17–22 was subjected to all three specifications. This data set was slightly different in that the third and fourth days were a Saturday and Sunday, respectively. All other data sets only considered weekdays. The results, which can be found in Table 3.2, indicated that the AR(1) model's parameter estimates were similar to the process estimated for the other five data sets. Specification I had an estimated AR parameter value of −0.92, significantly larger than any previous estimate. Specification II estimation shows a significant (at roughly a 95% confidence interval) AR(24) parameter value of 0.49, but the MA parameter is barely significant with 95%

TABLE 3.2 Estimation of three competing specifications on WEPCO unadjusted hourly load error: August 17–22

	d_1	$\hat{\sigma}d_1$	c_1	$\hat{\sigma}c_1$	$\hat{\sigma}_\eta$
SPECIFICATION I:					
$e'_t - e'_{t-24} = \dfrac{1}{1+d_1L'}\eta_t$	−0.92	0.03	—	—	42.99
SPECIFICATION II:					
$e'_t - e'_{t-1} = d_1L^{24}(e'_t - e'_{t-1}) + (1+c_1L')\eta_t$	0.49	0.08	0.19	0.09	40.38
SPECIFICATION III:					
$e'_t = \dfrac{1}{1+d_1L'}\eta_t$	−0.87	0.04	—	—	44.51

Note: e'_t represents the error in the unadjusted twenty-four hour load predictor.

confidence for the August 17–22 data set. As discussed earlier, estimation of Specification II on different WEPCO data sets supported different structural forms: AR(24), MA(1), or ARMA(24, 1).

It was our contention that an AR(1) model of the unadjusted hourly load error data was a simple and fruitful model to help predict hourly loads on the WEPCO system. The strength of the AR(1) model was the consistency of its parameter estimates for different types of days (e.g. weekdays vs. weekend) over different times of the year, and the model's ease of implementation.

4 WEPCO IMPLEMENTATION AND TESTING OF THE HOURLY LOAD PREDICTOR

The new hourly load predictor was tested over six days (November 6–9, 13–14, 1978). Comparisons were made between the existing WEPCO System (24-hour) Load Forecast (SLF) program and the first and second hour-ahead outputs of the hourly predictor. The observed standard deviations of the load predictions are shown in Table 3.3.

It is interesting to note from Table 3.4 that when the SLF program error is small (as on November 9) with near-zero mean value, the hourly predictor's performance is not much different from that of the system load forecast. This implies that the principal function performed by the hourly load predictor is the elimination of bias in the system load forecast.

An hour-by-hour comparison of the hourly predictor with actual WEPCO load on November 13, 1978 is given in Table 3.4.

Abbreviations used in Table 3.4 are:

SLF = System load forecast value for given hour
HP1 = Hourly predictor value, predicted one hour ago, for a given hour
HP2 = Hourly predictor value, predicted two hours ago, for a given hour

TABLE 3.3 Standard errors of hourly load predictors: six-day period

	Standard errors (MW)					
Prediction	Monday Nov. 6	Tuesday Nov. 7	Wednesday Nov. 8	Thursday Nov. 9	Monday Nov. 13	Tuesday Nov. 14
System load forecast (SLF)	60.45	27.59	27.3	21.65	103.86	100.45
Hourly predictor (One hour ahead)	36.00	25.84	22.9	24.24	29.12	32.49
Hourly predictor (Two hours ahead)	38.73	32.97	26.75	28.19	38.04	37.90

TABLE 3.4 WEPCO system load forecast and hourly predictor outputs for November 13, 1978

	Load and predictions (MW)				Prediction errors (MW)		
Hour	Actual	SLF	HP1	HP2	E_S	E_1	E_2
1	1493	1434	1524.2	1514.0	59	−31.2	−21.0
2	1430	1394	1454.0	1483.5	36	−24	−53.5
3	1422	1383	1420.6	1443.8	39	1.4	−21.8
4	1431	1371	1425.4	1430.9	60	5.6	0.1
5	1477	1424	1483.5	1477.9	53	−6.5	−0.9
6	1630	1583	1636.3	1642.2	47	−6.3	−12.2
7	2019	1930	1977.4	1983.5	89	41.6	35.5
8	2447	2341	2427.7	2388.8	106	19.3	58.2
9	2638	2537	2641.4	2621.9	101	−3.4	16.1
10	2713	2602	2702.8	2705.0	111	10.2	8.0
11	2769	2646	2757.3	2748.6	123	11.7	20.4
12	2786	2617	2739.2	2727.8	169	46.8	58.2
13	2675	2531	2697.1	2652.5	144	−22.1	22.5
14	2690	2546	2690.6	2709.8	144	−0.6	−19.8
15	2627	2478	2622.2	2623.0	149	4.8	4.0
16	2576	2462	2610.8	2606.3	114	−34.8	−30.3
17	2687	2505	2620.9	2653.6	182	66.1	33.4
18	2804	2672	2850.7	2789.5	132	−46.7	14.5
19	2732	2659	2792.9	2835.0	73	−60.9	−103.0
20	2662	2603	2679.9	2738.4	59	−17.9	−76.4
21	2596	2526	2587.0	2606.1	70	9	−10.1
22	2399	2358	2428.0	2420.6	41	−29	−21.6
23	2169	2118	2160.6	2187.9	51	8.4	−18.9
24	1929	1878	1928.9	1922.0	51	0.1	7.0

E_S = Error of the SLF value
E_1 = Error of the HP1 value
E_2 = Error of the HP2 value.

5 CONCLUDING REMARKS

The procedures and algorithms for hourly short-term prediction of electric power systems loads show the importance of iterative model specification, parameter examination, and re-specification until the key statistical tests are satisfied. The resulting predictor models, at least in the WEPCO application, show that relatively complex models which show inconsistent parameter estimates on independent data bases are reason for re-examination not only of the model structure, but of the data as well.

The relatively simple structure of Specification III yielded consistent parameter estimates for different types of days (e.g. weekdays vs. weekends) and for various seasons of the year. This model allowed WEPCO to account for 55% to 90% of the prediction error in hourly loads by making use of additional information inherent in the data series.

ACKNOWLEDGEMENTS

The author is indebted to Messrs Dale Ross, Robert Bischke, Robin Podmore, and Kent Wall. Any errors and omissions are the sole responsibility of the author.

REFERENCES

1. G. E. P. Box and G. M. Jenkins (1970). *Time Series Analysis: Forecasting and Control*, Holden-Day, San Francisco.
2. K. D. Wall (1976). 'Estimation of rational distributed lag structural form models.' *Annals Econ. Social Meas.*, **5,** No. 2, 53–63.
3. L. H. Fink, H. G. Kwatny, and J. P. McDonald (1969). 'Economic dispatch of generation via valve-point loading.' *IEEE Trans. on Power Apparatus and Systems*, **PAS-88,** No. 6, 805–811.
4. A. K. Cline (1974). 'Scalar- and planar-valued curve fitting using splines under tension.' *Communications of the ACM*, **17,** No. 4, 218–220.
5. A. K. Cline (1974) 'Algorithm 475, six subprograms for curve fitting using splines under tension [E2].' *Communications of the ACM*, **17,** No. 4, 220–223.
6. D. H. Budenaers (1973). 'Sequential short-term electric power demand forecasting.' *Proc. American Statistical Association*, **1973**.
7. K. C. Chen and P. R. Winters (1966). 'Forecasting peak demand for an electricity utility with a hybrid exponential model.' *Management Science*, **12,** No. 12.
8. W. R. Christiaanse (1970). 'Short-term load forecasting using general exponential smoothing.' *Trans. IEEE*, July, **1970**.
9. S. L. Corpening, N. D. Reppen, and R. J. Ringles (1973). 'Experience with weather-sensitive load models for short and long-term forecasting.' *Trans. IEEE*, November, **1973**.
10. E. D. Farmer and M. J. Potton (1968). 'Development of on-line load-prediction techniques with results from trials in the south-west region of the CEGB.' *Proc. IEEE*, **115,** No. 10.
11. F. D. Galiana and F. C. Schweppe (1971). 'A weather-dependent probabilistic model for short-term load forecasting.' *Trans. IEEE*, December, **1971**.
12. P. C. Gupta and K. Yamada (1972). 'Adaptive short-term forecasting of hourly loads using weather information.' *Trans. IEEE*, February, **1972**.
13. R. Pyne (1972). 'Bus load forecasting and its intended use in system security analysis.' Prepared for *Engineering Foundation Conference at BPA in Portland, Oregon, July 1972.*
14. J. Toyoda, *et al.* (1970). 'An application of state estimation to short-term load forecasting.' *Trans. IEEE*, **PAS-89,** No. 7, 1678–1686.
15. *Development and implementation of advanced automatic generation control software package*, U.S. Department of Energy Contract EC-77-01-2118, Systems Control, Inc., Palo Alto, California (L. H. Fink, U.S. Department of Energy Project Manager).
16. T. M. Athay, H. G. Kwatny and R. G. Smith (1978). 'Area load-frequency control.' *Proc. Sixteenth Annual Allerton Conference, Monticello, Illinois, October 4–6, 1978.*

Comparative Models for Electrical Load Forecasting
Edited by D. W. Bunn and E. D. Farmer

CHAPTER 4

Adaptive Short-term Forecasting of Hourly Loads Using Weather Information

P. C. Gupta

1 INTRODUCTION

The short-term load forecasting models are generally regression studies of load data and weather. More recently, time-series approaches have been tried such as Box–Jenkins or Kalman filters. This chapter will describe a different approach which has the following advantages;

1. The methodology takes both historical load data and weather information into account in forecasting hourly loads.
2. The forecasting models are adaptive in the sense that the model parameters are automatically corrected to keep track of the changing load conditions.
3. The methodology can be used effectively as an on-line software package in the utilities control computer system.
4. The methodology produces not only load forecasts but also the probable error in each forecast.

2 FORECASTING PROCEDURE

A highly simplified flow chart of the forecasting procedure is given in Figure 4.1. It combines two forecasting models: a stochastic model to relate the future loads to the past loads and a weather-load model to represent the effects of weather variables on the future loads. The models themselves are updated as the forecasts are produced. The procedure is discussed in more detail in the following sections.

2.1 Stochastic load model

The hourly system load is divided into three components:

$$Z(i, j) = T(i, j) + WC(i, j) + X(i, j) \tag{4.1}$$

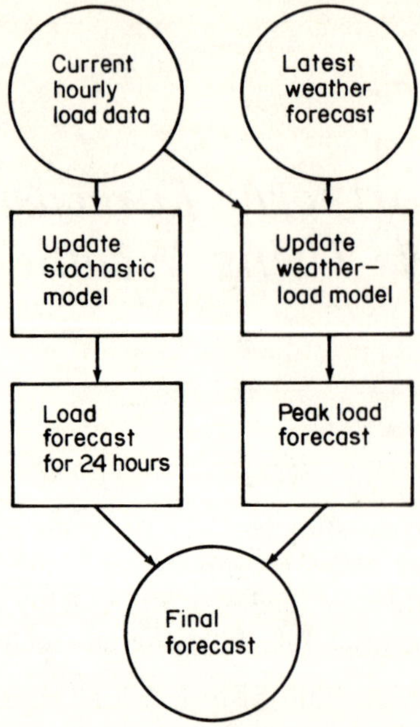

FIGURE 4.1 Simplified flowchart of the forecasting procedure

where

$Z(i, j)$ = system load measured in hourly MWh at hour j and day i
$T(i, j)$ = basic component of the load at hour j and day i
$WC(i, j)$ = weekly cycle component (day-of-the-week effect) at hour j and day i
$X(i, j)$ = residual component containing the effect of weather variations at hour j and day i.

The nature of these three components is, of course, very different. $T(i, j)$ can be considered more or less constant from day to day, $WC(i, j)$ is a slowly changing component signifying the weekly pattern of the hourly loads, and $X(i, j)$ is a rapidly changing component containing fast hour-to-hour variations in the load due to numerous random factors such as weather conditions experienced by major load centres. An autoregressive type of model is chosen for the random component:

$$\mathbf{X}(i) = A\mathbf{X}(i-1) + \mathbf{W}(i-1) \tag{4.2}$$

where

$$\mathbf{X}(i) = \begin{bmatrix} X(i, 1) \\ X(i, 2) \\ \vdots \\ X(i, 24) \end{bmatrix}: \quad \text{a } (24 \times 1) \text{ column vector of residual components for the } i\text{th day}$$

$A = (24 \times 24)$ matrix of coefficients

$\mathbf{W}(i-1)$ is a (24×1) column vector of model error terms.

An important characteristic of the model in equation (4.2) is that each element of $\mathbf{X}(i)$ is a linear function of all the elements of $\mathbf{X}(i-1)$. The error terms are assumed to be zero mean and statistically independent from day to day. An unknown (24×24) covariance matrix Q is defined such that

$$E(\mathbf{W}(i)\mathbf{W}(i)^{\mathrm{T}}) = Q \tag{4.3}$$

where $\mathbf{W}(i)^{\mathrm{T}}$ denotes the transpose of $\mathbf{W}(i)$. An equivalent form of equation (4.3) can be written for the (j, k)th element of Q as

$$Q(j, k) = E(W(i, j)W(i, k)) \tag{4.4}$$

where $W(i, j)$ is the model error term for hour j of the ith day.

2.2. Weather-load model

The forecasting procedure requires a weather-load model in which daily peak load (or the load at a fixed hour of the day) is represented in terms of weather variables. The model also contains terms corresponding to the basic load and the weekly pattern similar to those used in the stochastic model. The model is of the form

$$Y(i) = B(i) + S(i) + W(i) + \xi(i) \tag{4.5}$$

where

$Y(i)$ = peak load on the ith day
$B(i)$ = basic load component of the peak load on the ith day
$S(i)$ = weekly pattern component of the peak load on the ith day
$W(i)$ = weather-sensitive component of the peak load on the ith day
$\xi(i)$ = random component of the peak load on the ith day.

Strictly speaking, $B(i)$ and $S(i)$ should be defined as elements of $T(i, j)$ and $WC(i, j)$. However, better estimates can be expected to result if $B(i)$ and $S(i)$ are estimated independently of $T(i, j)$ and $WC(i, j)$, respectively.

The weather-sensitive component $W(i)$ is assumed to be a linear function of suitably transformed values of weather variables such as temperature and wind velocity observed at the major load centres. A general form of the model is given by

$$W(i)=\sum_{j=1}^{k} C_j WV_j(i) \tag{4.6}$$

where

k = number of weather variables in the model
$WV_j(i)$ = jth weather variable (or a transformed value of the variable) on the ith day
C_j = unknown jth coefficient of the model.

There will be seven values in the weekly pattern corresponding to each day of the week. Thus,

$$S(i)=\sum_{j=1}^{7} S_j P_j(i) \tag{4.7}$$

where

S_j = the jth value of the weekly pattern ($j=1$ corresponds to Sunday)

$$P_j(i)=\begin{cases} 1 & \text{if the } i\text{th day corresponds to the } j\text{th day of the week} \\ 0 & \text{otherwise.} \end{cases}$$

The weather variables selected are the average daily temperature and the average wind velocity. It is necessary to make a nonlinear transformation of the temperature variable in order to formulate a linear weather-load model of the type shown in equation (4.6). One of the many possible transformations is of the following form:

$$WV(i)=\begin{cases} TMP(i)-T_S & \text{if } TMP(i)>T_S \\ 0 & \text{if } T_W<TMP(i)<T_S \\ T_W-TMP(i) & \text{if } TMP(i)<T_W \end{cases} \tag{4.8}$$

where T_W and T_S are the fixed parameters of the transformation and $TMP(i)$ is the average temperature on the ith day.

2.3 Initialization procedures

Initialization procedures involve identification of the unknown parameters of the weather-load model and of the matrices A and Q of the stochastic model. In addition, the current values of $T(i, j)$, $WC(i, j)$, $X(i, j)$ of equation (4.1) and $B(i)$ and S_j of equations (4.5) and (4.7) are also required.

2.4 Identification of *A* and *Q*

The identification of A and Q is based on the residual component $\mathbf{X}(\cdot)$ of the historical load data. Since the residual component is not directly available, it is necessary to first estimate $T(i, j)$ and $WC(i, j)$ from the given load data. Each element of $T(i, j)$ is obtained by filtering the weekly pattern and the residual component from the load data:

$$T(i, j) = \frac{1}{7} \sum_{k=i-6}^{i} Z(k, j) \qquad \begin{array}{l} j = 1, 2, \ldots, 24 \\ i = 7, \ldots, N \end{array} \tag{4.9}$$

where N is the number of days for which load data is available.

The weekly pattern component $WC(i, j)$ is obtained by an exponential smoothing filter:

$$WC(i, j) = WC(i-7, j) + \alpha\{Z(i, j) - T(i, j) - WC(i-7, j)\} \qquad \begin{array}{l} j = 1, 2, \ldots, 24 \\ i = 8, 9, \ldots, N \end{array} \tag{4.10}$$

where α is a filter constant (0.2 to 0.5) and

$$WC(k, j) = Z(k, j) - T(7, j) \qquad k = 1, 2, \ldots, 7. \tag{4.11}$$

Hence, using equations (4.9)–(4.11), the residual components $\mathbf{X}(\cdot)$ can be estimated from the given load data. Let these be denoted by $\mathbf{X}(1), \mathbf{X}(2), \ldots$.

Post-multiplying equation (4.2) by $\mathbf{X}^{\mathrm{T}}(i)$ and taking the expected values gives—after some algebraic manipulation—the following:

$$Q = \Gamma_0 - A\Gamma_0 A^{\mathrm{T}} \tag{4.12}$$

where

$$\Gamma_0 = E[\mathbf{X}(i)\mathbf{X}^{\mathrm{T}}(I)]. \tag{4.13}$$

Similarly, post-multiplying equation (4.2) by $\mathbf{X}^{\mathrm{T}}(i-1)$ and taking the expected values gives

$$A = \Gamma_1 \Gamma_0^{-1} \tag{4.14}$$

where

$$\Gamma_1 = E[\mathbf{X}(i)\mathbf{X}^{\mathrm{T}}(i-1)]. \tag{4.15}$$

Equations (4.12) and (4.14) constitute the identification equations for identifying A and Q. Matrices Γ_0 and Γ_1 are estimated from the residual components $\mathbf{X}(\cdot)$ of the historical load data as follows:

$$\Gamma_0 = \frac{1}{N} \sum_{i=1}^{N} \mathbf{X}(i)\mathbf{X}^{\mathrm{T}}(i) \tag{4.16}$$

$$\Gamma_1 = \frac{1}{N-1} \sum_{i=2}^{N} \mathbf{X}(i)\mathbf{X}^{\mathrm{T}}(i-1). \tag{4.17}$$

2.5 Identification of the weather-load model

The unknown parameters of the weather-load model are identified by minimizing the exponentially weighted sum of squares of prediction errors.

That is to say, we minimize

$$E=\sum_{n=1}^{N}\left[Y(n)-B(n)-\sum_{i=1}^{7}S_iP_i(n)-\sum_{i=1}^{k}C_iWV_i(n)\right]^2\beta^{N-n} \qquad (4.18)$$

subject to the estimation of $B(n)$ by an exponential smoothing scheme,

$$B(n)=B(n-1)+\alpha\left[Y(n-1)-\sum_{i=1}^{7}S_iP_i(n-1)-\sum_{i=1}^{k}C_iWV_i(n-1)-B(n-1)\right] \qquad (4.19)$$

where β is a constant between 0 and 1. Constant α is also assumed known. Experimentally it is seen that if α is selected to be less than 0.1, the estimation of $B(n)$ loses its adaptability and if it is chosen to be more than 0.3, the estimation becomes unstable. Hence the value of α should lie within these limits. Estimation of $B(n)$ by exponential smoothing as shown in equation (4.19) takes into account 'the accumulative effect of demand.' This effect can be observed in the peak loads during summer; once the habit of using air conditioners has been formed, the peak load does not generally decrease in proportion to the temperatures.

Equation (4.19) can be written as

$$B(n)=\bar{Y}(n-1)-\sum_{i=1}^{k}C_i\overline{WV}_i(n-1)-\sum_{i=1}^{7}S_i\bar{P}_i(n-1) \qquad (4.20)$$

where

$$\bar{Y}(n-1)=\alpha Y(n-1)+(1-\alpha)\alpha Y(n-2)+\cdots$$
$$=\alpha Y(n-1)+(1-\alpha)\bar{Y}(n-2) \qquad (4.21)$$

$$\overline{WV}_i(n-1)=\alpha WV_i(n-1)+(1-\alpha)\alpha\overline{WV}_i(n-2)+\cdots \qquad (4.22)$$
$$=\alpha WV_i(n-1)+(1-\alpha)WV_i(n-2) \qquad (4.22)$$

$$\bar{P}_i(n-1)=\alpha P_i(n-1)+(1-\alpha)\alpha P_i(n-2)+\cdots$$
$$=\alpha P_i(n-1)+(1-\alpha)\bar{P}_i(n-2). \qquad (4.23)$$

Hence, equation (4.18) becomes

$$E=\sum_{n=1}^{N}\left[Y(n)-\bar{Y}(n-1)-\sum_{i=1}^{7}S_i(P_i(n)-\bar{P}_i(n-1))\right.$$
$$\left.-\sum_{i=1}^{k}C_i(WV_i(n)-\overline{WV}_i(n-1))\right]^2\beta^{N-n}. \qquad (4.24)$$

If the initial values of $\bar{Y}(n)$, $\overline{WV}_i(n)$ $(i=1,2,\ldots,k)$, and $\bar{P}_i(n)$ $(i=1,2,\ldots,7)$ are known, the subsequent values can be estimated using equations (4.21)–(4.23) and the unknown parameters S_i and C_i can be determined by a least-squares method. For computational economy, a recursive estimator is used. Let

$$\Delta Y(n) = Y(n) - \bar{Y}(n-1) \tag{4.25}$$

$$a^{\mathrm{T}} = (S_1, S_2, \ldots, S_7, C_1, C_2, \ldots, C_k) \tag{4.26}$$

$$\begin{aligned}\phi^{\mathrm{T}}(n) = [&P_1(n) - \bar{P}_1(n-1), P_2(n) - \bar{P}_2(n-1), \ldots, \\ &WV_1(n) - \overline{WV}_1(n-1), WV_2(n) - \overline{WV}_2(n-1), \ldots,]\end{aligned} \tag{4.27}$$

$\hat{a}(N)$ = optimum estimate of the coefficient vector of a at the Nth iteration. Then (as shown in the Appendix),

$$a(N) = a(N-1) + P(N)\phi(N)[\Delta Y(N) - \phi^{\mathrm{T}}(N)a(N-1)] \tag{4.28}$$

where $P(N)$ is a newly defined variable,

$$\begin{aligned}P(N) = \frac{1}{\beta^P}(N-1) - \frac{1}{\beta^2}P(N-1)\phi(N) \\ \times\left[1 + \phi^{\mathrm{T}}(N)\frac{1}{\beta}P(N-1)\phi(N)\right]^{-1}\phi^{\mathrm{T}}(N)P(N-1).\end{aligned} \tag{4.29}$$

To perform the identification process by using equations (4.28) and (4.29), we require initial values of $\bar{Y}(\cdot)$, $\overline{WV}_i(\cdot)$, $\bar{P}_i(\cdot)$, $\hat{a}(\cdot)$, and $P(\cdot)$. In order to obtain these, the data base is divided into three parts as shown in Figure 4.2. Part A is used for determining $\bar{Y}_0$, $\overline{WV}_{i0}$, $\bar{P}_{i0}$, and the initial values of $Y(n)$, $WV_i(n)$, and $P_i(n)$. The initial values are determined by taking the simple averages

$$\bar{Y}_0 = \frac{1}{m}\sum_{n=1}^{m} Y(n) \tag{4.30}$$

$$\overline{WV}_{i0} = \frac{1}{m}\sum_{n=1}^{m} WV_i(n) \qquad i = 1, 2, \ldots, k \tag{4.31}$$

$$\bar{P}_{i0} = \frac{1}{m}\sum_{n=1}^{m} P_i(n) \qquad i = 1, 2, \ldots, 7 \tag{4.32}$$

where m is the total number of days included in Part A. Part B is used for

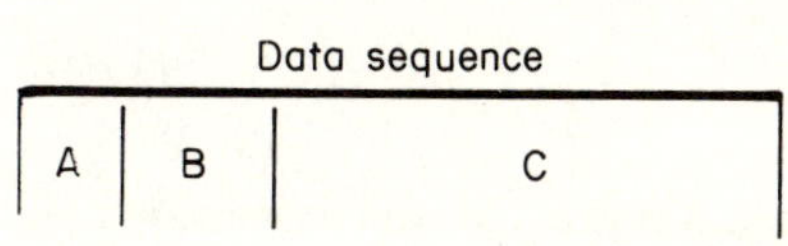

FIGURE 4.2 Division of the historical database into three parts for identification of the weather-load model

the determination of initial values $\hat{a}(0)$ and $P(0)$. Using equations (4.21)–(4.23) and (4.27), the values of $\overline{WV}_i(n)$ and $\phi(n)$ are obtained. Then $\hat{a}(0)$ and $P(0)$ are estimated from the following equations:

$$P(0)=\left[\sum_{j\notin \text{Part } B} \phi^{\mathrm{T}}(j)\phi(j)\right]^{-1} \tag{4.33}$$

$$\hat{a}(0)=P(0)\left[\sum_{j\in \text{Part } B} \phi(j)\,\Delta Y(j)\right]. \tag{4.34}$$

Part C of the data is used for updating $P(\cdot)$ and $\hat{a}(\cdot)$ recursively and the final values are stored for producing peak load forecasts with the current weather-load data.

2.6 Adaptive forecasting procedure

The load forecast for the $(i+1)$th day, $\hat{Z}(i+1,j)$, $j=1,2,\ldots,24$, is made on the basis of measurements $Z(i,j)$, $j=1,2,\ldots,24$, and the weather forecast $\tilde{WV}_j(i+1)$, $j=1,2,\ldots,k$, in the following steps (refer to Figure 4.1).

Update $T(i,j)$ and $WC(i,j)$ The basic component $T(i,j)$, $j=1,2,\ldots,24$ is updated according to equation (4.9) and the weekly pattern component $WC(i,j)$ is updated according to equation (4.10). The residual component $\mathbf{X}(i)$ is then determined as follows:

$$X(i,j)=Z(i,j)-T(i,j)-WC(i,j) \qquad j=1,2,\ldots,24. \tag{4.35}$$

Update A and Q Matrices A and Q are updated by re-estimating the covariance matrices Γ_0 and Γ_1 using equations (4.16) and (4.17). This type of re-estimation is, however, computationally inefficient since a major part of the calculation is repeated from one update to the other. A more efficient algorithm for updating the estimates of Γ_0 and Γ_1 is given by equations (4.36) and (4.37):

$$\Gamma_0^i=\Gamma_0^{i-1}+\frac{1}{i}[\mathbf{X}(i)\mathbf{X}^{\mathrm{T}}(i)-\Gamma_0^{i-1}] \tag{4.36}$$

$$\Gamma_1^i=\Gamma_1^{i-1}+\frac{1}{i-1}[\mathbf{X}(i)\mathbf{X}^{\mathrm{T}}(i-1)-\Gamma_1^{i-1}] \tag{4.37}$$

where Γ_0^i and Γ_1^i are the estimates of the covariance matrices Γ_0 and Γ_1 obtained by processing the data of the ith day. Equations (4.36) and (4.37) can be easily derived from equations (4.16) and (4.17).

The algorithm given above is very efficient since the updating of the covariance matrices is based on only the previous estimates and the new product terms $\mathbf{X}(i)\mathbf{X}^{\mathrm{T}}(i)$ and $\mathbf{X}(i)\mathbf{X}^{\mathrm{T}}(i-1)$.

The new estimates of A and Q are computed using updated estimates of Γ_0 and Γ_1 in equations (4.12) and (4.14).

Forecast based on stochastic model The forecast of the next 24 values of the residual component $\hat{\mathbf{X}}(i+1)$ is produced by using equation (4.2) and assuming the forecast of model errors $\mathbf{W}(i+1)$ to be the espected value of $\mathbf{W}(i+1)$, i.e. zero:

$$\hat{\mathbf{X}}(i+1) = A\mathbf{X}(i). \tag{4.38}$$

The load forecast based on the stochastic model is:

$$\hat{Z}_S(i+1, j) = T(i, j) + WC(i+1, j) + \hat{X}(i+1, j) \qquad j = 1, 2, \ldots, 24. \tag{4.39}$$

In the forecasting equation (4.39) the basic component is assumed to be unchanged from the ith to the $(i+1)$th day. The variance of the forecast $\hat{Z}_S(i+1, j)$ is given by the variance of the jth model error term, i.e. $Q(j, j)$:

$$\sigma_S^2(j) = Q(j, j). \tag{4.40}$$

Update weather-load model The actual peak load for the ith day, $Y(i)$, is determined from the load data:

$$Y(i) = \operatorname*{Max}_j [Z(i, j)]. \tag{4.41}$$

By using equations (4.21), (4.22), (4.23), (4.25), (4.28), and (4.29), the parameters of the weather-load model $\hat{a}(i)$ are updated. The values of the weather variables in the updating procedure can either be the weather forecast made on the previous day or the actual values observed on the ith day.

Peak load forecast based on weather forecast The peak load forecast for the $(i+1)$th day is calculated as follows:

$$\hat{Y}(i+1) = B(i+1) + S(i+1) + \sum_{j=1}^{k} C_j W\tilde{V}_j(i+1) \tag{4.42}$$

where $W\tilde{V}_j(i+1)$ is the forecast of the jth weather variable for the $(i+1)$th day.

The variance of $\hat{Y}(i+1)$, σ_y^2, is computed by updating the variance estimate based on the observed peak-load forecast error of the ith day, i.e.

$$\sigma_y^2 = \sigma_y^2 + \frac{1}{1}[\{Y(i) - \hat{Y}(i)\}^2 - \sigma_y^2]. \tag{4.43}$$

Composite load forecast At this point, we have a forecast $\hat{Z}_S(i+1), j)$, $j = 1, 2, \ldots, 24$, produced by the stochastic model with the associated covariance matrix Q and a peak-load forecast $Y(i+1)$ with variance σ_y^2

based on the weather forecast. The composite load forecast should be an optimal combination of the two forecasts. If the maximum value of $\hat{Z}_S(i+1, j)$, $j = 1, 2, \ldots, 24$, is the same as $\hat{Y}(i+1)$, the composite forecast is given by $\hat{Z}_S(i+1, j)$, $j = 1, 2, \ldots, 24$. However, the weather forecast brings extra information about the future loads and hence, in most cases, $\hat{Y}(i+1)$ and the maximum of $\hat{Z}_S(i+1, j)$ are different.

The procedure for combining two stochastic estimates is quite well known in the modern control literature. It basically amounts to obtaining a suitable weighted average of the two estimates. The equations for computing a composite load forecast $\hat{Z}(i+1, j)$ from $\hat{Z}_S(i+1, j)$ and $\hat{Y}(i+1)$ are the following:

$$\hat{Z}(i+1, j) = \hat{Z}_S(i+1, j) + \frac{[Y(i+1) - \hat{Z}_{SP}]Q(j, p)}{Q(p, p) + \sigma_y^2} \tag{4.44}$$

where

$$\hat{Z}_{SP} = \max_j \{\hat{Z}_S(i+1, j)\} \tag{4.45}$$

and p is the hour of the day associated with $\hat{Z}_{SP}$. The forecast variance, $\hat{\sigma}_j^2$, is reduced from its original value since extra information in the form of the peak-load forecast has been used in the composite forecast:

$$\hat{\sigma}_j^2 = Q(j, j) - \frac{Q^2(j, p)}{Q(p, p) + \sigma_y^2} \qquad j = 1, 2, \ldots, 24. \tag{4.46}$$

3 PERFORMANCE OF THE PROCEDURE

The forecasting methodology has been run on actual load data obtained from an electric utility. For this purpose, the following data were obtained

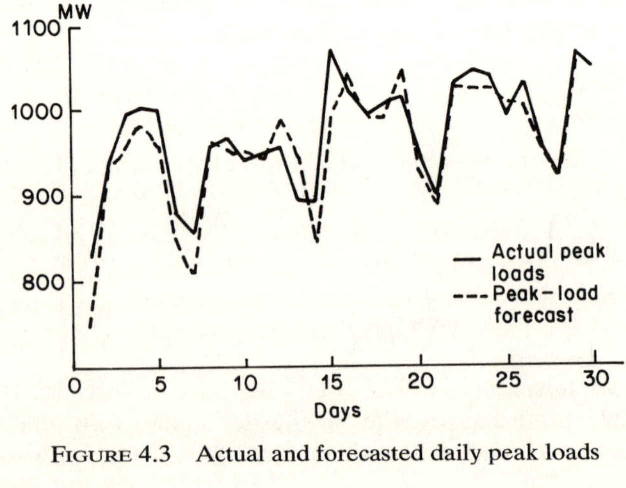

FIGURE 4.3 Actual and forecasted daily peak loads

TABLE 4.1 Typical forecast produced at 0 hour with and without weather forecast consideration

Hour	Forecast produced by model (MW)	Peak-load forecast based on weather forecast (MW)	Composite load forecast (MW)	Actual (MW)	Error (MW)
1	615		612	602	−10
2	568		565	543	−22
3	539		534	524	−10
4	520		515	499	−16
5	516		509	498	−11
6	511		504	511	7
7	542		536	528	−8
8	600		587	597	10
9	706		683	700	17
10	804		783	783	0
11	867		826	839	13
12	874	823	852	837	−15
13	853		827	815	−12
14	827		798	803	5
15	798		767	773	6
16	756		722	721	−1
17	760		731	726	−5
18	781		751	762	11
19	803		772	753	−19
20	792		762	733	−29
21	804		774	759	−15
22	803		795	818	23
23	751		746	766	20
24	693		688	676	−12

from the utility:

1. Hourly load data for a period of one and a half years.
2. Historical weather data recorded by a weather station at the prominent load centre of the utility for the same period as load data.

The actual weather data were used since the weather forecast record was not readily obtainable. Specifically, the data consisted of 3-hourly weather observations recorded on magnetic tape. Another data tape containing daily peak loads and the daily average values of temperature and wind velocity was generated from the original load and weather data tapes.

The results of a simulation of a period of one month are given in Figure 4.3 where the actual peak loads are plotted, along with their forecasts as made one day ahead.

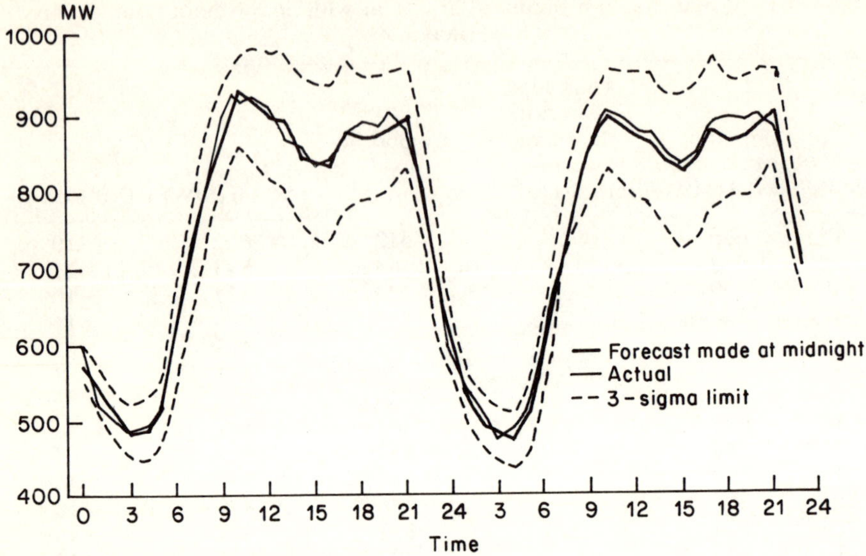

FIGURE 4.4 Hourly load forecast and 3-σ limits compared with the actual loads that materialized

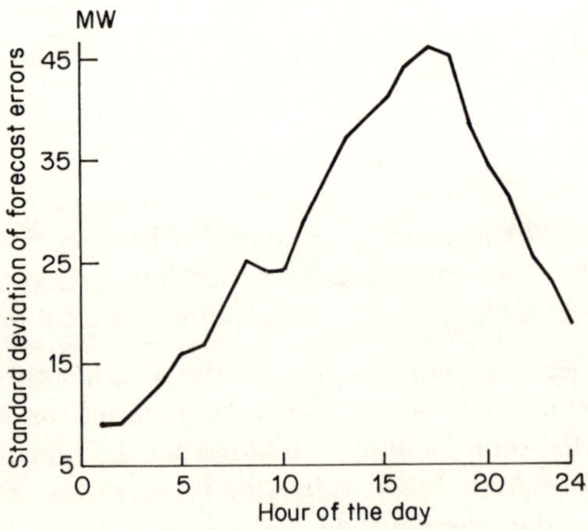

FIGURE 4.5 Standard deviations of the forecast errors as a function of the hour of the day

Table 4.1 summarizes a typical forecast and Figure 4.4 illustrates the forecast accuracy by comparing the actual load data with the forecasts produced at the beginning of each day.

The forecast errors over the total simulation period are summarized in Figure 4.5, where the standard deviations of the forecast errors are plotted as a function of the hour of the day. Since the average load during those six months was approximately 1000 MW, it can be seen that the largest standard deviation is about 4.5% of the load. It is also observed from Figure 4.5 that the forecast errors are definitely a function of the hour of the day, with the largest errors occurring at 5 p.m. and the smallest at 1 a.m. for the system studied.

APPENDIX DERIVATION OF RECURSIVE EQUATIONS (4.28) AND (4.29) FOR UPDATING THE PEAK-LOAD FORECASTING MODEL

Rewriting (4.24), we get

$$E=\sum_{n=1}^{N}[\Delta Y(n)-a^{\mathrm{T}}\phi(n)]^2\beta^{N-n}. \tag{A4.1}$$

Let $\hat{a}(N)$ be the optimum estimate which minimizes E. Then $\hat{a}(N)$ will satisfy the equation

$$\left.\frac{\partial E}{\partial a}\right|_{a=\hat{a}(N)}=0 \tag{A4.2}$$

or,

$$\hat{a}(N)=\left[\sum_{n=1}^{N}\phi(n)\beta^{N-n}\phi^{\mathrm{T}}(n)\right]^{-1}\left[\sum_{n=1}^{N}\phi(n)\,\Delta Y(n)\beta^{N-n}\right]. \tag{A4.3}$$

Let

$$P(N)=\left[\sum_{n=1}^{N}\phi(n)\beta^{N-n}\phi^{\mathrm{T}}(n)\right]^{-1} \tag{A4.4}$$

then

$$\hat{a}(N)=P(N)\sum_{n=1}^{N}\phi(n)\,\Delta Y(n)\beta^{N-n} \tag{A4.5}$$

or,

$$\begin{aligned}P(N)^{-1}\hat{a}(N)&=\sum_{n=1}^{N}\phi(n)\,\Delta Y(n)\beta^{N-n}\\&=\beta\sum_{n=1}^{N-1}\phi(n)\,\Delta Y(n)\beta^{N-1-n}+\phi(N)\,\Delta Y(N)\\&=\beta P(N-1)^{-1}\hat{a}(N-1)+\phi(N)\,\Delta Y(N).\end{aligned} \tag{A4.6}$$

Equation (A4.4) can be rewritten in recursive form as follows:

$$P(N)^{-1}=\beta P(N-1)^{-1}+\phi(N)\phi(N)^{\mathrm{T}}. \tag{A4.7}$$

Hence,

$$P(N)^{-1}\hat{a}(N) = (P(N)^{-1} - \phi(N)\phi(N)^{\mathrm{T}})\hat{a}(n-1) + \phi(N)\,\Delta Y(N)$$
$$= P(N)^{-1}\hat{a}(N-1) + \phi(N)(\Delta Y(N) - \phi(N)^{\mathrm{T}}\hat{a}(N-1)). \tag{A4.8}$$

Pre-multiplying (A4.8) by $P(N)$, we get equation (4.28):

$$\hat{a}(N) = \hat{a}(N-1) + P(N)\phi(N)(\Delta Y(N) - \phi(N)^{\mathrm{T}}\hat{a}(N-1)).$$

From equation (A4.7)

$$P(N) = (\beta P(N-1)^{-1} + \phi(N)\phi(N)^{\mathrm{T}})^{-1}. \tag{A4.9}$$

Equation (4.29) can be derived by using equation (A4.9) and the well-known matrix inversion lemma,

$$(A + BCB^{\mathrm{T}})^{-1} = A^{-1} - A^{-1}B(C^{-1} + B^{\mathrm{T}}A^{-1}B)^{-1}B^{\mathrm{T}}A^{-1}. \tag{A4.10}$$

Comparative Models for Electrical Load Forecasting
Edited by D. W. Bunn and E. D. Farmer

CHAPTER 5

Load Forecasting for Scheduling Generation on a Large Interconnected System

A. B. Baker*

1 INTRODUCTION

Consumer demand for electricity from the CEGB's National Grid System exhibits large daily and seasonal variations; a ratio of approximately 4:1 exists between winter peak demand and a summer night demand. The total installed capacity of the generating plant on the system is determined by the planning criterion that a suitably small risk should exist for failure to meet the highest expected demand in winter. It therefore follows that considerable savings can be achieved by arranging to run only just sufficient generating plant to meet consumer demand at any time during the remainder of the year. Further savings are attainable by utilizing the most economical generating plant available at any time and scheduling this plant in the most cost-effective manner within the constraints of the plant and system performances. Additionally savings can be achieved by minimizing the amount of spare plant which must be run to cover contingencies, such as breakdown or load forecasting errors.

These cost optimizations are achieved by the national coordination of power transfer on the transmission system. Although in the past thorough cost optimization was sought at particularly critical periods of the day, known as costing periods, located in general at the turning points of the daily load curve, modern data processing techniques are now allowing the full optimization to progress to a growing fraction of each day.

Demand forecasting techniques originally developed to service the scheduling process allied to costing periods have been retained and augmented by curve fitting techniques to provide load predictions at times between costing periods to service the computer optimized scheduling.

* The views expressed in this paper are those of the author and not necessarily those of the CEGB.

2 THE SHORT-TERM FORECASTING MODEL

No description of a forecasting technique would be complete without an exposition of the historic data series upon which it is based. Data banks of demand and weather have been accumulated daily on spinning-disk storage of an IBM 370 computer and now cover a period of 12 years. The demand data bank takes the form of daily records, each record subdivided into observations of electrical power demand integrated over 30-minute intervals for each of the CEGB Grid Control Areas together with a CEGB total. These demand records are stored in two forms: a met demand record and an unrestricted demand record. The unrestricted demand record is derived from the met demand corrected for effects such as load management demand relief, sustained operation outside of frequency control limits (±0.2 Hz), load shedding by low voltage operation, and consumer disconnection. It is this unrestricted demand record, which is the best available detailed record of consumer demand as seen at the generator terminals, which is referred to when demand, or load, is mentioned hereafter.

The weather data bank is similarly compiled on a daily basis with records of processed weather parameters for five periods of each day from observations made, at selected weather stations, by the Meteorological Office. Selection of suitable weather station sites is determined principally by proximity to the major conurbations within the Grid Control Areas. A weighted average of these processed weather observations is also stored as the CEGB weather record. The form of these processed weather parameters will be described below.

Identification of the causal relationship between weather and electricity demand and the formulation of load forecasting models is effected by multiple linear regression analysis. A purpose built regression analysis program has been written to access the demand and weather data banks providing a rapid analytical aid.

Two types of short-term load forecasting models are in current use, having the following outline formulations:

(a) Demand = Weather dependent load + basic load + day of week adjustment + error
(b) Demand = Weather dependent load + basic load + function of earlier load + day of week adjustment + error.

Type (a) model is clearly a purely additive model, in which each of the disaggregated components may be forecast, involving demand and weather for a single costing period from common day grouped data series. Type (b) model uses, as a basis for prediction, a measure of a recently experienced load on the current day adding corrections for forecast changes in weather and predictable trends. As a result of these fundamental differences type (a) models are best suited to forecasts with lead-times in excess of 12 hours running up to 3 days, using forecast weather as an input, but may be used with lead-times as long as a few weeks making use of probabilistic weather bases. Type (b) models, as would be expected, are found to perform best with very short lead-times, typically 3 to 6 hours.

3 WEATHER-DEMAND RELATIONSHIPS

As early as the late 1930s weather was identified as being a major source of short-term variation in electricity demand in the UK. This causal relationship acts through a variety of consumer applications including space heating, water heating, refrigeration, air-conditioning, and lighting. Space heating load is affected by air temperature, both current values and recent past temperature history, and windspeed. Water heating, refrigeration, and air conditioning demand are influenced by air temperature and lighting demand, during the hours of daylight, is dependent on the natural sky illumination.

It is this short-term weather/demand relationship which is modelled in the regression analysis; the longer-term seasonal effects are represented by the time series approach to the basic load component.

Four derived weather parameters have been identified to model the weather induced electricity demand variation. These are as follows.

3.1 Average temperature, TO_r

For each weather station this is the average of the 4 spot hourly temperature observations up to time r. That is,

$$TO_r = \tfrac{1}{4} \sum_{n=0}^{3} T_{r-n}.$$

3.2 Effective temperature, TE_r

This is an exponentially smoothed form of TO for common r values:

$$TE_r = aTO_r + (1-a)TE_{r-24}$$

where TE_{r-24} is the effective temperature at time r on the previous day.

The sensitivity of the smoothing coefficient, a, close to its optimum value has been found to be noncritical, no significant difference being detected between values of 0.4 and 0.6. For simplicity a value of 0.5 is currently in use.

The form of the effective temperature function implies a time lag in the response of heating appliances within buildings to changes in external temperature. This is consistent with the physical heat transfer processes involved.

3.3 Cooling power of the wind, CP_r

This is an empirical combination of temperature and windspeed, W_r:

$$CP_r = \begin{cases} W_r^{1/2}(18.3 - TO_r) & \text{for} \quad TO_r < 18.3\,°\text{C} \\ 0 & \text{for} \quad TO_r \geqslant 18.3\,°\text{C}. \end{cases}$$

This formulation is found to cope adequately with draught-induced load variation without the additional information of wind direction.

3.4 Effective illumination of the sky, EI_r

The Meteorological Services of the UK do not make routine measurements or forecasts of daylight illumination. In order to formulate a predictive model of electrical lighting load, relationships between observable meteorological parameters, such as visibility, cloud cover, and precipitation, have had to be established. These relationships were derived from observations made at Kew over an extended time period and take the following form:

$$EI_r = MI_r - ID_r$$

where

EI_r is Effective Illumination of the sky at time r

MI_r is Maximum Illumination of the sky (under conditions of perfect visibility and no cloud)

ID_r is Illumination Deficit which is a function of visibility, number, and types of cloud layers and amount and type of precipitation.

For a single cloud layer with an apparent transmission factor of f the illumination deficit is simply expressed by

$$ID = 100 \log_{10}\left(\frac{1}{f}\right).$$

When more than one layer of cloud is present with less than perfect visibility (when the visibility is treated as another cloud layer) the deficits are combined to produce an overall illumination deficit given by

$$ID = 100 \log_{10}\left[\left\{\sum_{i=1}^{n} \text{Antilog}_{10}\left(\frac{ID_i}{100}\right)\right\} - (n-1)\right]^{-1}.$$

The Kew observations have been analysed to yield values of apparent transmission factor for 10 categories of cloud type, each of 8 oktas of sky occlusion, in 3 bands of cloud base height, with 5 categories of precipitation, and all conditions of visibility.

4 MODELLING THE WEATHER-DEPENDENT LOAD

Judgment, stemming from experience, is essential when selecting historic data to be analysed by regression techniques in order that the derived model may not only best fit the past, but is likely to provide a sound representation of the future.

A compromise in the size of the data set must be achieved from considerations of statistical confidence, which increases with the size of the data set, and the ability of the model to reflect changing consumer habits which requires successive modelling on small data sets.

The characteristics of the CEGB load are such that, at present, each year may be modelled in two segments determined by the clock time regime applied in the UK, these being the Greenwich Mean Time period (late October through to late March) and the British Summer Time period (the remainder of the year). Days of the week are grouped into sets with common characteristics for analysis. Weekdays (Monday to Friday) exhibit similarities conditioned by the social timetable associated with the 5 day work cycle. Both Saturdays and Sundays have unique load characteristics which require each to be separately analysed.

Historic data series are scrutinized to exclude discontinuities which would significantly corrupt the regression analysis. Included in this category are Public Holidays, block industrial holidays, and days affected by industrial strikes. Days adjacent to these calendar events may also be affected to a lesser extent and are examined carefully.

In order to improve statistical confidence in the analysis several years of data may be overlaid. This technique is particularly useful when analysing Saturdays or Sundays which are subject to data paucity.

The formulations of the weather dependent loads differ in detail between models. Two typical examples are given below.

4.1 Model type (a) weather/load representation

$$\text{Weather dependent load} = f(TE_r) + \beta EI_r + \gamma CP_r$$

where

$$f(TE_r) = \begin{cases} \alpha_1 TE_r + \alpha_2 TE_r^2 & \text{for} \quad TE_r < \dfrac{-\alpha_1}{2\alpha_2}, \\[2ex] \dfrac{-\alpha_1^2}{4\alpha_2} & \text{for} \quad TE_r \geqslant \dfrac{-\alpha_1}{2\alpha_2}. \end{cases}$$

Thus $f(TE_r)$ is a discontinuous parabolic function of temperature taking on a constant value for temperatures above the turning point. This function is used to model the nonlinear response of load with temperature, with the magnitude of the response reducing with increasing temperature, having a zero rate of change for temperatures above a level given by the turning point. Typically this level is found to be about 20 °C.

This form of temperature/load modelling was determined from data obtained during the summer of 1976 when the UK experienced summer temperatures expected to occur once per century. The zero response of load to change in temperature at high temperatures is thought to have occurred as a result of near perfect equilibrium, at the time, between residual negatively temperature responsive load (in particular water heating) and positively temperature responsive load (refrigeration and air conditioning). Recent experience during short spells of high temperature suggest that this equilibrium may have been disturbed since 1976 by the growth of food freezer and air conditioner sales.

Linear representations of load response with respect to effective illumination and cooling power of the wind are found to be adequate.

As an example of the magnitude of the coefficients in this form of weather dependent load representation the following values are quoted from a model currently in use to forecast winter weekday lunchtime peak demands. These coefficients operate on forecast weather parameters for 1200 hours:

$$\alpha_1 = -545 \text{ MW/°C}$$

$$\alpha_2 = 12.75 \text{ MW/°C}^2$$

$$\beta = -30.0 \text{ MW/unit of illumination}$$

$$\gamma = 48.8 \text{ MW/unit of cooling power.}$$

Typical ranges of variation of EI and CP during winter are 100 and 70 respectively.

4.2 Model type (b) weather/load representation

$$\text{Weather dependent load} = \alpha_1 TO_{r_1} + \beta_1 EI_{r_1} + \gamma_1 CP_{r_1} + \alpha_2 TO_{r_2} + \beta_2 EI_{r_2} + \gamma CP_{r_2}.$$

As a result of the formulation of type (b) models which associates load at two periods within a day, r_1 and r_2, it is found necessary for the weather-dependent component to contain weather parameters for both periods to account for weather changes.

A typical example of this form of weather-load model can be seen in a current winter weekday model to forecast the lunchtime peak load having experienced the breakfast time peak load:

$$\text{Weather dependent load} = 61.5TO_{09} + 28.1CP_{09} - 90.9TO_{12} - 19.2EI_{12}$$

where the subscripts 09 and 12 refer to 0900 hour and 1200 hour weather parameters.

It is seen in this example that terms in EI_{09} and CP_{12} are missing from the formulation. This has resulted from nonsignificant estimates being obtained from the regression analysis for coefficients on these parameters. Also to be seen in this example is the opposition of the signs associated with the coefficients on TO_{09} and TO_{12} demonstrating how the two weather sets compensate for weather-load changes between the two periods.

The use of the temperature functions TO in this formulation, as distinct from TE used in type (a) models, is explained by the fact that much of the lagging effect of load on changing temperature is already represented by the function of an earlier load within the total demand model.

5 REPRESENTATION OF DAY OF WEEK ADJUSTMENTS

Day of week variation is estimated in the regression analysis by dummy variables D_i, given by a 5×5 unit matrix, as follows:

$$\text{Day of week adjustment} = \sum_{i=1}^{5} \delta_i D_i$$

for a typical weekday model (Monday to Friday) where δ_i are the estimated coefficients.

6 MODELLING OF THE BASIC LOAD COMPONENT

Both type (a) and type (b) models have a component which is described as a basic load. This component exhibits seasonality and is also a function of other exogenous influences such as economic growth, manufacturing production, and price competition with other energy sources, in particular gas.

For the purpose of regression analysis the basic load is represented by a time series in the form of a polynomial in day number, a function incrementing by 1 with the calendar. In order to ensure that the polynomial adequately fits the long-term trend and seasonal components of the load without encroaching upon the short-term variation induced by weather and day of week, the lowest order of polynomial consistent with these requirements is used. In practice a polynomial to the 4th power of day number is found to be adequate.

It must be stressed that this simplistic representation of basic load is used only during the analytical stage of modelling load in order that the remaining coefficients may be efficiently estimated by the multiple regression process.

7 THE EARLIER LOAD FUNCTION IN TYPE (b) MODELS

It has been found that an adequate representation of this term in type (b) models is given by kL_{r_1}, where k is the regression coefficient and L_{r_1} the load at the earlier time r_1.

In general it is found that the value of k is in the range 0.5 to 1.0. Models associating period r_1 and r_2 separated by 3 to 4 hours have values close to the top of the range, whereas models with lead-times of 8 to 10 hours have values of k towards the bottom of the range. These findings are consistent with expectations based upon the autocorrelation properties of the daily load curve.

8 APPLICATION OF THE REGRESSION MODELS IN THE PREDICTIVE MODE

A computer program has been written for the sole purpose of applying the regression models of load to produce short-term load forecasts based upon weather forecasts. This program accepts weather forecasts, coded in a computer compatible format, issued by Meteorological Office Regional Weather Centres to an agreed timetable. These weather forecasts apply to the major conurbations of the CEGB Grid Control Areas, for consistency with the historic weather data used to formulate the load models. Recent weather forecasts are stored by the program, principally to allow type (b) models to be evaluated before actual weather observations become available. As many as 56 load models may be stored within the program allowing more than one model to be used for forecasting a particular period. Currently up to four models are used for a single forecast period; the forecaster is thus able to critically assess the short run performance of individual models as an aid to his decision making.

In addition to the forecast weather requirements of the models, inputs of earlier loads and basic loads are required. Recently experienced earlier loads are quantitatively estimated by reference to the output of an on-line demand validation and very short-term predictor running on a process control computer in real-time, which takes as its input the telemetered outputs of the generators connected to the system and the power transfers associated with the CEGB system.

In order to produce predictions of basic loads a tracking procedure is maintained for each model. The original historic demand and weather series upon which each model was evaluated is arithmetically reprocessed using the fitted regression models to produce an historic series of basic load + model error given by:

Type (a) models: Load − weather dependent load − day of week adjustment.

Type (b) models: Load–weather dependent load–day of week adjustment –function of earlier load

As more recent load and weather data accrues in the season during which the model is being applied for forecasting, further points are overlaid on the plots which serve as graphical aids to assist the trend prediction.

The graphical tracking method has been found to yield better results than strict mathematical fitting or smoothing routines by virtue of the fact that the experienced forecaster is able to make use of all of the information available and not just a mathematical condensation.

9 TREATMENT OF SPECIAL DAYS

The reader will remember that discontinuities in the historic demand series are excluded from the regression analysis from which the load models are developed. In order that the continuous requirement for load forecasts may be met, techniques have been developed which rely upon the load models derived from these 'laundered' series. This approach has been necessitated by the failure to model the sparse data provided by the special days themselves.

The technique starts by examination of the daily load curves of historic occurrences of the special days, the technique being particularly suitable to Public Holidays and adjacent nonstandard days, in order to determine similarities with close standard days. Usually the main Public Holidays are found to resemble a Sunday in load shape, although particular days, for example Good Friday, have Saturday characteristics. The postulate that load shape similarities implies load mix similarities and hence load/weather response similarities, is made. Thus by using the weather-dependent load coefficients of the best shape fit standard day models the load histories of these special days may be analysed and projections of the basic load derived and used together with forecast weather to provide load forecasts in the usual way.

10 PERFORMANCE OF THE FORECASTING METHOD

As expected the errors in forecast load provided by the techniques described above increase with lead-time. The errors also exhibit seasonality with the largest errors occurring in spring and autumn. This effect is thought to be the result of the volatility of weather and the ensuing increase in weather forecasting uncertainty together with the increase in electricity consumers' volatility at these times.

During winter and summer forecasting errors of 1.4% r.m.s. are obtained for lead-times of 3 to 4 hours, and 2.5% r.m.s. for lead-times of 24 to 36 hours. These errors increase to 1.8% and 3.0% respectively during spring and autumn.

11 FORECASTING THE DAILY LOAD CURVE

In order to obtain forecasts of load at times of the day not serviced by load forecasts derived from the above techniques, an interpolation program has been written.

An historic day, which is considered will represent the current day in respect of shape-influencing characteristics, in particular sky illumination, day of week, and calendar proximity, is selected for the program to access load data for this day from either of two data banks. One data bank comprises integrated half-hourly loads and has been described above. A spot load data bank provides the source data for a more detailed load curve prediction.

The selected load curve is constrained to pass through turning point load levels, determined by the load forecasts input from the regression model techniques by a curve stretching process.

12 EFFECTS OF TELEVISION PROGRAMMING ON LOAD

The effects of television audience release on electricity load has been observed for a number of years. Popular television programmes inhibit normal domestic electricity consuming habits during the course of the programmes, only to release this activity at commercial breaks and at the finish of the programme. Large and rapid changes in electricity demand occur at these release times and are known as 'TV pickups'. In order to schedule generating plant to anticipate these events forecasts of the times and magnitudes of these TV pickups are required. Although regression models based upon variables such as audience rating, program classification, and time of day have been investigated, heuristic techniques relying upon experienced forecasters making use of detailed program and load histories and maintaining close liaison with the television broadcasting authorities are currently employed.

13 CONCLUSIONS

A practical scheme for forecasting electrical loads on a large interconnected system has been described in some detail. These techniques have been developed, and are employed, in the operational environment of the National Control Centre of the CEGB transmission system and as a result have benefited from the rapid feedback inherent in the real-time world.

REFERENCES

1. M. Davies (1958). *The relationship between weather and electricity demand*, IEE Monograph No. 3145, October 1958.

2. A. B. Baker (1977). 'Forecasting for the operational requirements of the CEGB National Grid System.' (*Proc. Public Utilities Forecasting Conference, Bowness-on-Windermere, March 1977.*)
3. E. D. Farmer, W. D. Laing, A. M. Adatia, A. B. Baker, and D. W. Bunn (1981). 'The Development and Implementation of an Interactive Demand Validation and Prediction Facility.' *Seventh PSCC, Lausanne, July 1981.*
4. A. M. Adatia, A. B. Baker, E. D. Farmer, W. D. Laing, and D. G. Robertson (1982). 'Interactive Load Forecasting and Reserve Allocation.' *CIGRE Conf. 32–23, September 1982.*
5. D. W. Bunn and J. P. Seigal (1981). *Tracking Audience Release During Breaks in Television Programming*, University of Oxford Department of Engineering Science Report OUEL 1353/81, April 1981.

Comparative Models for Electrical Load Forecasting
Edited by D. W. Bunn and E. D. Farmer

CHAPTER 6

Time Series Methods for Predicting the CEGB Demand

W. D. Laing

1 INTRODUCTION

In a climate of increasing energy costs and uncertainty over the availability and costs of different sources of energy, there is a potential need to centralize and refine the optimization of the loading of the electrical generation plant for the whole of the CEGB power system. The optimization is concerned with meeting the demands of the consumer as economically as possible, with due regard to operational system security constraints. This operation is determined by the plant on the bars at the current time, and the planned unit commitment schedule, but, even within these restrictions, costs can be reduced by an economic dispatch of the available generation. This must take account of the expected variations in the consumer demand as well as the capability and flexibility of plant to respond to the changes requested, and sufficient reserve must be allocated to meet contingencies such as errors in the forecasted demands, plant shortfalls, and operational security requirements.[1] In a power system which has immediate response gas turbine or pumped storage plant all of this reserve may not be held on the thermal plant.[2] Indeed, gas turbine and pumped storage plant may provide the most economic way of meeting sudden, but expected, short-term demand variations, such as a television pickup, and must be included in the dispatch calculation. Thus there is a fundamental requirement for the provision of a demand forecast which covers the dispatching period in sufficient detail to model the minute-to-minute variations which might arise. Furthermore, if on-line real-time data is available for regular updates of the forecasted demand curve, dynamic revision of the dispatch can be expected to further reduce costs and increase operational security.

For longer lead times, covering up to one day ahead, there is also a need for accurate forecasts of the demand, although not necessarily with the same detail as those covering the dispatching period. Such a requirement arises from the role played by the pumped storage plant. Whether such plant is operated in a reserve allocation role, or as part of the dispatch, there is a daily cycle in which the water used must, on average, be replenished in readiness for the next day. The longer term forecast will aid the calculation of the optimum utilization of the water remaining before the next pumping period.

A project which investigates the economic benefits of a centralized dispatch of available generation is currently under development within the CEGB. A computer, situated at the Board's National Control Centre, receives on-line power system data from the six area control centres, and facilities for the input of off-line information, required in the current development, are being provided. The demand prediction facility forms an integral part of this software.

This chapter describes the methods which have been used to provide demand forecasts covering the two periods given above. Some of the problems which have arisen in establishing a robust, fully automated demand prediction facility are described, and possible solutions are discussed. Some results which reflect the current performance of the facility are presented.

2 ON-LINE DATA

The basic data required by the prediction facility consist of databanks of historical demands which are stored and updated in a common on-line database. These data are derived from real-time data telemetered to the National Control Centre, the raw data being filtered to remove any errors and then integrated over one- and thirty-minute intervals to provide the averaged demand values entered into the databanks. The data validation programs use simple time series methods to validate the raw metered data but, at present, have no cross validity checks with other available metered records. This has led to bad data being passed into the databanks with the subsequent deterioration in the performance of the demand prediction facility. The poor performance may persist for several weeks because of the underlying assumptions of the time series prediction algorithms, and bad data may be transmitted through the databank if interpolation routines are called upon to estimate missing data resulting from failure in the telemetry or computer systems. Improvements which use other available data (both raw and calculated) are currently under investigation, and it is hoped that these will reduce the quantity of bad data entering the database to a manageable level, even if some good values are also rejected. An on-line interactive display facility is also under development in which the demand history can be examined, and erroneous values identified and corrected manually. This provides the final stage in the validation process, with manual inspection of the stored demands being carried out during slack periods, or when the performance of the predictor indicates the presence of uncharacteristic demand data.

Databanks are stored for both national and grid control area demands covering periods of about two weeks of one-minute averages and about three years of half-hour averages. The national demand is simply the sum of the indvidual area demands which are corrected to a system frequency of 50 Hz using a load-frequency sensitivity of 2% per Hz.

The telemetry data rate ensures that between 3 and 5 spot demands are available each minute depending upon the area, although system frequency data is received at a rate of 5 values per second. The data are asynchronous between areas and are not time-stamped until they are received by the on-line computer for entry into the database.

3 TIME SERIES FORECASTING METHODS

Two methods for time series forecasting have been implemented for the prediction of electrical demand for periods up to 24 hours ahead. Forecasts are estimated for each area and for the national system, although only the latter prediction is used directly by the dispatching algorithm. The predictions derived from the two methods are combined to provide a single record of demand forecasts covering the next twenty-four hours at one-minute intervals. Other data from off-line sources may also be available, and, if input, are used in the combinatorial algorithm. Such data would include forecasts calculated off-line using weather information and entered either manually or via an inter-computer link. Information relating to the expected demand variations for television pickups and load management functions would also be entered manually and superimposed on the estimated demand curve.

3.1 Spectral decomposition predictor

Historical records of either national of area demands exhibit certain regular patterns which do, on average, repeat themselves from day to day, week to week, and year to year. If these components are extracted from the demand records, then the residual values represent perturbations about the seasonal average due to variations in the local conditions, such as weather, which affect consumer demand. Thus

$$D(d, t) = D_s(d, t) + y(d, t) \tag{6.1}$$

where

d is the day of week
t is the time of day
D_s is the seasonal average component
y is the residual load.

Taking discrete times, t_n $(1 \leq n \leq N)$, and putting

$$y_m(t_n) = y(d - 7m, t_n) \qquad m = 1, \ldots, M \tag{6.2}$$

where

M is the number of weeks' history
N is the number of time intervals spanned

the spectral decomposition algorithm[3] expands the residual loads in the form

$$\hat{y}_m(t_n) = \sum_{k=1}^{K} c_{mk}\phi_k(t_n) \tag{6.3}$$

to minimize the mean-square residual

$$s^2 = (MN)^{-1} \sum_{m=1}^{M} \sum_{n=1}^{N} [\hat{y}_m(t_n) - y_m(t_n)]^2. \tag{6.4}$$

The minimum error is given by

$$\hat{Y} = Y\Phi^{\mathrm{T}}\Phi \tag{6.5}$$

where

$$(Y)_{mn} = y_m(t_n)$$

$$(\hat{Y})_{mn} = \hat{y}_m(t_n)$$

$$\Phi^{\mathrm{T}} = (\boldsymbol{\phi}_1, \ldots, \boldsymbol{\phi}_K)$$

where

$\boldsymbol{\phi}_k$ are the eigenvectors corresponding to the K largest eigenvalues of the covariance matrix

$$R = M^{-1}Y^{\mathrm{T}}Y. \tag{6.6}$$

The elements c_{mk} and $\phi_k(t_n)$ are given by equation (6.5):

$$\begin{aligned} c_{mk} &= (Y\Phi^{\mathrm{T}})_{mk} \\ \phi_k(t_n) &= (\boldsymbol{\phi}_k)_n. \end{aligned} \tag{6.7}$$

Equation (6.3) represents an approximation, with K eigenfunctions, of the $(M \times N)$ sample function Y of residual loads over a period of N intervals. The prediction problem is that having identified the characteristic modes (6.3) for this period, and, in addition, having available demands for the current day covering the period, t_n $(1 \leq n \leq N_H)$, what is the best estimate of the demand for the period t_n $(N_H < n \leq N)$?

If a characteristic mode expansion of the form

$$\hat{y}_{0n} = \hat{y}_0(t_n) = \sum_{k=1}^{K} c_{0k}\phi_k(t_n) \tag{6.8}$$

is proposed, the coefficients, c_{0k}, can be estimated by minimizing the mean square error

$$s_0^2 = N_H^{-1} \sum_{n=1}^{N_H} [\hat{y}_{0n} - y_{0n}]^2 \tag{6.9}$$

to give

$$\mathbf{c}_0 = (\Phi_0 \Phi_0^T)^{-1} \Phi_0 \mathbf{y}_0 \tag{6.10}$$

where

$\mathbf{c}_0$ is the K-vector of coefficients c_{0k}
$\mathbf{y}_0$ is the N_H-vector of residuals y_{0n}
Φ_0 is a $(K \times N_H)$ partition of the matrix $\Phi = (\Phi_0, \Phi_1)$.

The predicted residuals are calculated by extrapolating the eigenfunctions expansion (6.8) over the period t_n $(N_H < n \leq N)$, to give

$$\hat{\mathbf{y}}_0 = \Phi_1^T \mathbf{c}_0 \tag{6.11}$$

where

$$\hat{y}_{0n} = (\hat{\mathbf{y}}_0)_n$$

is the estimated residual for time t.

It is clear from the above analysis that, since the eigenfunctions span the whole interval $\{t_n : 1 \leq n \leq N\}$, the predicted residuals depend upon the value chosen for N, whereas it would be desirable that an estimate for a particular lead time was independent of N. To overcome this problem a single time-step predictor, having a parallel with the Kalman filter,[4] has been implemented. In this algorithm the eigenvectors of the reduced covariance matrix, consisting of the $(N_H + 1) \times (N_H + 1)$ partition of N_H historical time-steps together with the row and column corresponding to the first predicted time-step, are calculated, and the first residual is estimated using (6.11), where the partition, Φ_1, is a column vector. The procedure is repeated for the subsequent prediction lead times by augmenting the reduced covariance matrix with the row and column corresponding to the next lead time, adding each predicted residual to the vector of historical residuals as it is calculated, and applying (6.11) as before.

In practice, since the number of weeks' history, M, is much smaller than the number of half-hour periods, N, a more efficient algorithm is given by using the alternative covariance matrix S:

$$S = M^{-1} Y Y^T. \tag{6.12}$$

The nonzero eigenvalues of S are the same as those of R, and the corresponding eigenvectors of R can be obtained from those of S by premultiplication by Y^T. For the previous analysis to be valid the eigenvectors are normalized so that the sum of squared elements is unity. With the single time-step method, described above, the matrix S is updated for each lead time by adding the appropriate column product from the matrix of residuals, Y. Further computational savings are achieved by using the eigenvectors obtained from the previous step as the initial approximation in the iterative scheme used for determining the dominant eigenvalues for the next time-step.

The forecasted demands are calculated by adding the seasonal component according to equation (6.1).

A number of different methods are being examined for separating the seasonal and residual components of equation (6.1). The first method examined set up the seasonal component by averaging over the M weeks covering the period of the historical load curves:

$$D_s(d, t) = M^{-1} \sum_{m=1}^{M} D(d-7m, t). \tag{6.13}$$

While this gives a fast, easily automated procedure, the performance of the predictor is poor at those times of the year when the shape of the load curve changes from week to week, such as during the spring and autumn. This is because the seasonal component lags behind the current date, and is also constant for the M-week period, and hence does not follow the weekly changes in load pattern which occur at these times of the year. A method in which the seasonal component is established from load curves of previous years is currently under investigation.

Another algorithm which gave some improvement in the performance of the predictor over the first method was a spectral decomposition of differenced demands

$$y(d, t) = D(d, t) - D(d, t-T) \tag{6.14}$$

where T is the sampling interval.

Again the time series $y_m(t_n)$ is easily set up, but the performance of the predictor still suffers from the same problems as the first method, although to a lesser degree.

A ratio deviation residual, calculated by

$$y(d, t) = \frac{D(d, t) - D_s(d, t)}{D_s(d, t)} \tag{6.15}$$

was also tested with the spectral decomposition algorithm described above and gave, on average, less accurate forecasts, although the difference was small compared with the prediction errors observed.

3.2 Exponential weighted moving average predictor

A prediction algorithm has been implemented which uses the databank of one-minute averages to generate a sequence of demand residuals which are processed by a second-order low pass filter.[5] The residuals are calculated by either (6.1) or (6.15), where the seasonal component is taken to be the load for the same time and day of the previous week. Although this approach does not generate a genuine seasonal component, it is the best approximation that can be achieved at the present time. Disk limitations prevent the storage of more than about two weeks of one-minute load averages, although the planned enhancements to the on-line computer will permit the investigation of improved methods for estimating the seasonal demand component.

If the impules response function of the filter takes the form

$$h(\tau)=\frac{1}{T_2-T_1}\left\{\frac{T_2+T_3}{T_1}\mathrm{e}^{-\tau/T_1}-\frac{T_1+T_3}{T_2}\mathrm{e}^{-\tau/T_2}\right\} \tag{6.16}$$

where T_1, T_2, and T_3 are three time constants, the extrapolated residual $\hat{y}(t+\theta)$ is given by the convolution integral

$$\hat{y}(t+\theta)=\int_0^\infty y(t-\tau)h(\tau)\,\mathrm{d}\tau \tag{6.17}$$

where θ is the lead time.

The frequency response function, $H(f)$, is given by

$$H(f)=\int_0^\infty h(\tau)\mathrm{e}^{-2\pi\mathrm{i}f\tau}\,\mathrm{d}\tau$$

which for the filter (6.16) becomes

$$H(p)=\frac{1+p(T_1+T_2+T_3)}{1+p(T_1+T_2)+p^2T_1T_2} \tag{6.18}$$

where

$$p=2\pi\mathrm{i}f.$$

It is clear that this filter extrapolates a linear trend for a lead time T_3 without error since, if

$$y(t)=at+b$$

where a and b are constants, then

$$\hat{y}(t+T_3)=y(t+T_3).$$

If the residual is a combination of a linear trend and a noise component, $n(t)$,

$$y(t)=at+b+n(t)$$

then the optimum value of the time constant T_3 can be calculated by minimizing the mean-square prediction error

$$s^2=E[(\hat{y}(t+\theta)-y(t+\theta))^2]$$

where E is the expectation operator.

This reduces to

$$s^2 = a^2(\theta - T_3)^2 + \int_0^\infty \int_0^\infty h(\tau)h(\tau')R(\tau - \tau')\,d\tau\,d\tau'$$

where $R(\tau)$ is the autocorrelation function of $n(t)$.

If $n(t)$ is a white noise process, its autocorrelation function is given by

$$R(\tau) = \sigma^2 \delta(\tau)$$

where σ^2 represents the noise power per unit bandwidth and $\delta(\tau)$ is the Kronecker delta function, and the mean-square error becomes

$$s^2 = a^2(\theta - T_3)^2 + \sigma^2 \int_0^\infty h^2(\tau)\,d\tau.$$

Minimizing the mean-square error gives

$$T_3 = \frac{\theta - \dfrac{\sigma^2}{2a^2T_1T_2}}{1 + \dfrac{\sigma^2}{2a^2T_1T_2(T_1+T_2)}} \tag{6.19}$$

which shows that the effective lead time, T_3, of the filter decreases as the ratio of the noise power, σ^2, to the energy in the trend term, a^2, increases.

The impulse response function (6.16) can be rearranged as

$$h(\tau) = h_1(\tau) + \frac{T_1 + T_3}{T_2 - T_1}\{h_1(\tau) - h_2(\tau)\}$$

where $h_1(\tau)$ and $h_2(\tau)$ are two single exponential responses

$$h_i(\tau) = \frac{1}{T_i} e^{-\tau/T_i} \qquad i = 1, 2.$$

Substituting for the reduced lead time (6.19) gives

$$h(\tau) = h_1(\tau) + W\frac{(T_1 + \theta)}{(T_2 - T_1)}\{h_1(\tau) - h_2(\tau)\} \tag{6.20}$$

where θ is the lead time and W is a weighting factor given by

$$W = \frac{1 + \dfrac{\theta}{T_1} - \dfrac{\sigma^2}{2a^2T_1^2(T_1+T_2)}}{\left\{1 + \dfrac{\sigma^2}{2a^2T_1T_2(T_1+T_2)}\right\}\left\{1 + \dfrac{\theta}{T_1}\right\}}. \tag{6.21}$$

It is clear that the weighting factor is unity in the absence of noise and is less than unity if random fluctuations are present. Furthermore, the weighting factor tends to one as the lead time, θ, increases.

In practice the double exponential filter, equation (6.16), is applied in its sampled data form. If T is the sampling interval the extrapolated residual (6.17) becomes

$$\hat{y}(t+\theta) = \sum_{k=0}^{\infty} h_k y(t-kT) \tag{6.22}$$

where the discrete impulse response coefficients, h_k are given by

$$h_k = T\int_{-1/2T}^{1/2T} H(f)e^{2\pi ifkT}\,df \tag{6.23}$$

where $H(f)$ is the frequency response function which can be evaluated by using the difference analogue to the differential equation described by (6.18)

$$T_1T_2\frac{d^2y}{dt^2}+(T_1+T_2)\frac{dy}{dt}+y=(T_1+T_2+T_3)\frac{dx}{dt}+x.$$

Using backward differences, a linear difference equation

$$\begin{aligned}(T_1+T)(T_2+T)y(t)-[T_1(T_2+T)+T_2(T_1+T)]y(t-T)+T_1T_2y(t-2T)\\ =T(T+T_1+T_2+T_3)x(t)-T(T_1+T_2+T_3)x(t-T)\end{aligned}$$

is obtained.

Taking Fourier transforms

$$Y(f)=H(f)X(f)$$

where

$$Y(f)=\int_{-\infty}^{\infty} y(t)e^{-2\pi ift}\,dt \qquad X(f)=\int_{-\infty}^{\infty} x(t)e^{-2\pi ft}\,dt$$

the frequency response function becomes

$$H(f)=\frac{p[(T+T_1+T_2+T_3)p-(T_1+T_2+T_3)]T}{(T_1+T)(T_2+T)p^2-[(T_2+T)T_1+(T_1+T)T_2]p+T_1T_2} \tag{6.24}$$

where

$$p=e^{2\pi ifT}.$$

The characteristic equation for the discrete system is

$$(T_1+T)(T_2+T)p^2-[T_1(T_2+T)+T_2(T_1+T)]p+T_1T_2=0$$

which has roots given by

$$p_i=\frac{T_i}{T+T_i} \qquad i=1, 2.$$

The discrete impulse coefficients, h_k, now become the contour integrals

$$h_k = \frac{1}{2\pi i} \oint_{|p|=1} \frac{(\alpha p + \beta)p^k}{(p-p_1)(p-p_2)} \, dp$$

where

$$\alpha = \frac{(T+T_1+T_2+T_3)T}{(T+T_1)(T+T_2)} \qquad \beta = -\frac{(T_1+T_2+T_3)T}{(T+T_1)(T+T_2)}.$$

The contour integral is evaluated by the Cauchy residue theorem, giving

$$h_k = \frac{T_2+T_3}{T_2-T_1}(1-p_1)p_1^k - \frac{T_1+T_3}{T_2-T_1}(1-p_2)p_2^k$$

which can be rearranged to give

$$h_k = h_{ik} + \frac{T_1+T_3}{T_2-T_1}(h_{1k}-h_{2k}) \tag{6.25}$$

where

$$h_{ik} = (1-p_i)p_i^k \qquad i = 1, 2.$$

Equation (6.25) is the sampled data analogue of the continuous impulse response function (6.16).

Finally, the extrapolated residuals are calculated using (6.22) to give

$$\hat{y}(t+\theta) = y_1(t) + \frac{T_1+T_3}{T_2-T_1}(y_1(t)-y_2(t))$$

where

$$y_i(t) = (1-p_i) \sum_{k=0}^{\infty} p_i^k y(t-kT) \qquad i = 1, 2$$

are exponentially weighted moving averages (ewmas).

The two ewmas can, more conveniently, be calculated by the recurrence relations

$$y_i(t) = p_i y_i(t-T) + (1-p_i)y(t) \qquad i = 1, 2. \tag{6.26}$$

Substituting for the reduced lead time, T_3, from (6.19) the predicted residual becomes

$$\hat{y}(t+nT) = y_1(t) + W\frac{(T_1+nT)}{(T_2-T_1)}(y_1(t)-y_2(t) \tag{6.27}$$

where nT is the lead time, θ, and W is given by (6.21).

Several other weighting functions have also been examined using histories of logged demand residuals. These varied from constant values of 1 or 0, which extrapolate a linear trend without error or reduce to a single ewma predictor, respectively, to complex functions involving the lead time nT and the time constants T_1 and T_2. An empirical function which gave reasonable results is

$$W = \exp\left[-(nT/T_0)^2\right] \tag{6.28}$$

where the decay time constant, T_0, was chosen following off-line tests. The residual extrapolation (6.27) was also modified to

$$y(t+nT) = W\left(y_1(t) + \frac{(T_1+nT)}{(T_2-T_1)}(y_1(t)-y_2(t))\right). \tag{6.29}$$

The weighting function (6.28) approaches unity for short lead times, so that (6.29) extrapolates along a linear trend for the immediate future. For longer lead times, however, the weighting factor tends to zero, so that the residual load also tends to zero and the demand forecasts approach the reference seasonal curve.

4 COMBINING FORECASTS

The two algorithms described above generate forecasts of one- and thirty-minute demand averages for the national and grid control area systems for the next twenty-four hours. Selfconsistent forecasts for each area and for national are required which present a smooth interface between the one- and thirty-minute forecasts, and which, additionally, incorporate any off-line forecasts which may also be available. A program is currently under development to implement these tasks and uses the prediction error statistics, accumulated by the on-line prediction programs, in the combinatorial algorithm. At present, the one-minute forecasts and logged errors are integrated to form thirty-minute averages corresponding to the periods of the half-hour predictor, and a correction factor is calculated for each half-hour interval from

$$\Delta D_{30} = \frac{s_{30}^{-2} D_{30} + s_1^{-2} D_1}{s_{30}^{-2} + s_1^{-2}} - D_{30} \tag{6.30}$$

where

D_{30}, s_{30}^2 are the thirty-minute forecast and mean-square error
D_1, s_1^2 are the corresponding values for the thirty-minute averages based upon the one-minute forecasts
ΔD_{30} is the half-hour average demand forecast correction.

The half-hour correction, which is an average for the thirty-minute period, is converted into adjustments to the one-minute demand forecast record by interpolation as described in the Appendix. The algorithm generates one-minute sub-interval values from the half-hour correction factors such that the original averages are recovered if the sub-interval values are integrated over the half-hour. Assigning the correction to the half-hour midpoint, and interpolating, introduces errors in the recovery of the half-hour averages if curvature is present.

Manually entered demand variations for TV pickups and load management reductions are superimposed on the estimated load curve, and the twenty-four hour prediction is stored in the database for display and for input to the dispatch algorithm. Confidence limits for the predicted load curve are estimated based on previous prediction errors and the errors logged for the separate sources of demand forecasts.

It is proposed to incorporate off-line, manually entered, forecasts by calculating load curve adjustment values in a way that is similar to the combination method described above. The off-line forecasts are half-hour integrated demands, based upon weather information, which cover the critical turning points in the daily load curve. Up to six such values, together with their estimated time of occurrence, are available with lead times of several hours, although revisions may be made as the critical time is approached in the light of changes in the weather pattern. A suitable display is being developed for easy entry of the off-line forecasts into the demand prediction facility. An interactive demand forecasting facility was developed for the National Control Centre's on-line monitoring and display computer but, because of programming problems, was not successfully completed.[5]

5 RESULTS

Tables 6.1 and 6.2 show a summary of some of the result which have been logged for the two demand forecasting algorithms described above. The results have been calculated from the demand prediction error statistics which are dumped each day to monitor the performance of the demand

TABLE 6.1 CEGB half-hour average demand prediction errors (spectral decomposition)

Lead (half-hour)	RMS error (%)
1	2.7
2	3.0
4	3.2
12	3.7
24	3.8
48	3.9

TABLE 6.2 CEGB one-minute average demand prediction errors (double exponential)

Lead (minute)	RMS error (%)
1	0.4
2	0.6
5	0.9
10	1.1
30	2.1
60	3.2
120	4.1
240	4.8
720	4.9
1440	4.9

prediction facility. Those days on which the performance of the predictor was affected by abnormal situations, such as Bank Holidays, or by bad data entering the demand databanks, have been excluded since methods of overcoming these problem areas have not yet been integrated. Some ideas for providing forecasts covering these periods are discussed in the next section, or, in the case of data validation, were discussed earlier. The tables show that the one-minute predictor gives smaller root-mean-square prediction errors for lead times up to about one hour, but that for longer lead times the half-hour predictor is more accurate. The reason for the relatively large errors for short lead times with the spectral decomposition algorithm is not known and subsequent results have shown the errors for lead times of one and two half-hours reduced to almost half of the values shown.

The results presented are based upon the performance of the predictors over a few weeks in an on-line real-time environment. A simulation has been developed whereby the effects of parameter modifications can be fully tested by replaying the prediction facility over the same common set of demand data dumped from the database, although the simulator was not available for a retrospective analysis of the demand data used for the results shown in the tables.

The regular daily dumps of prediction error statistics also include the largest prediction error for that day. For the autumn period monitored, the worst predictions usually occurred at the time of the onset of darkness in the early evening, the errors being several times larger than the root-mean-square averages shown in the tables.

The results for the spectral decomposition algorithm were based on using the previous five weeks to establish the seasonal component and the residual covariance matrix, the seasonal component being derived from (6.13). The historical period consisted of 24 half-hours for a prediction of up to 48 half-hour time steps ahead, and only the eigenvectors corresponding to the two largest eigenvalues of the covariance matrix were used in the spectral representation. Tests were made, using the off-line simulator, on the effect of modifying the time periods and parameters used in the on-line program. Only small variations in prediction accuracy were observed over a wide range of values for the historical period, N_H (values from 12 to 48 were tried), and only marginal improvements were achieved by using a larger number of characteristic modes, K, in the spectral decomposition model. Different values for the weeks of history, M, were more difficult to test because of the occurrence of abnormal periods at fairly frequent intervals during the year. Exhaustive tests using different numbers of weeks will be done when methods which remove the effects of such special days are implemented.

In the ewma one-minute prediction algorithm, values used for the two time constants, T_1 and T_2, were 1 and 5 minutes, respectively, and the values used for the weighting function decay parameter, T_0, was 10 minutes. Other values for the time constants T_1 and T_2 have been tested and it was found that only marginal differences in the performance of the predictor occurred with values of the ratio of T_2 to T_1 in the range 3 to 5, with values of T_1 in the range 0.5 to 3. The longer lead time forecast errors for the ewma predictor show how closely last week's load curve models the load for the current week and, for the autumn period chosen, reflect the substantial weekly variations in the shape of the load curve which may occur at this time of year.

6 PROBLEM AREAS

The previous sections have described the need for a demand prediction facility, its data requirements, the methods which have been implemented or are under development, and some of the results reflecting the current performance of the facility. In the implementation of a fully automated real-time facility, consideration must be given to abnormal situations which may arise, some of which may be known in advance with others occurring apparently at random. Solutions to these problems must be found if the facility is to be regarded as reliable and robust, a necessary objective when the economics of the dispatch calculation are considered, with its dependence on accurate, regular demand forecasts.

The first problem encountered was that of British Summer Time (BST) and its effect on the daily load cycle. As a first step towards solving this problem, it was agreed that the Dispatch Project computer should run to Greenwich Mean Time (GMT) with all time-stamped on-line data received and entered into the database to this time base. This has the advantage that time would be monotonic increasing, so that data would not have to be thrown away at the end of the BST period, or extra data added at the start of BST, simply to maintain a constant number of daily demand averages. A problem which remains, however, is the effect of the changeover on the consumer's use of electrical power. To a GMT time base, the load pattern governed by the regular daytime consumer activities is altered by one hour by the changeover, affecting the morning load rise and end-of-the-working-day peak. Other regular daily load patterns, such as the time-of-darkness lighting peak, the night storage load, which may be significant at these times of year, and the night minimum trough, are not directly affected. Thus, a forecasting algorithm which relies upon repeatable load patterns from week to week to calculate new demand estimates is seriously affected by the introduction of BST. One solution to this problem is to use data from previous years, in addition to the load history already used, by identifying the same day of week relative to the GMT/BST changeover as a reference day for that year, and retrieving historical periods relative to this reference. The use of data from previous years will be investigated in the expectation that the problems encountered with the GMT/BST changes will be reduced.

Another problem, which should also be aided by using data from previous years, is that of predicting the time-of-darkness lighting peak. This was identified as a problem area in previous work,[5] and is verified by the results obtained so far and discussed earlier. During the spring and autumn, the time of the onset of darkness changes by up to 13 minutes in one week, so that using data from a few weeks ago introduces errors in the estimate of the seasonal load curve. Adjustment functions, both empirical and theoretical, which modify the estimated seasonal curve have been tested, but have not been successful.

A further problem is that of special days such as Bank Holidays and their neighbouring days. Information relating to these special days will be manually entered via an on-line display facility, and databanks will be set up to contain dumped load curves from the demand databanks when they are no longer required on-line. They would be retrieved for the prediction of demands for future special days. A system in which the special day load curves were generated off-line and loaded into a databank has been tested in the demand prediction facility developed for the National Control Centre's on-line monitoring and display computer.[5]

With the developments outlined above, and with improvements to the data validation algorithm, it is hoped that the root-mean-square errors can be reduced to the larger of 0.5% or 200 MW for half-an-hour ahead and 2% or 600 MW for four hours ahead on a national demand which has an annual range of between 10 and 45 gigawatts. These would be acceptable targets to be achieved by the on-line demand prediction facility and approach those prediction errors, for the longer lead time, achieved by the weather-weighted regression models in use by the CEGB.[6] It should be re-stated that the on-line predictor would make use of the forecasts provided by the weather-weighted models to achieve this accuracy.

7 CONCLUSIONS

A demand prediction facility has been implemented as part of a project investigating central dispatch and, within the limitations of the present state of development, has performed with reasonable accuracy. Some problem areas have been identified and methods of overcoming these difficulties have been discussed in the expectation that the performance should be improved to provide a reliable, robust, on-line, real-time demand prediction facility.

ACKNOWLEDGEMENT

The work was carried out at the Central Electricity Research Laboratories and is published by permission of the Central Electricity Generating Board.

REFERENCES

1. A. M. Adatia, A. B. Baker, E. D. Farmer, W. D. Laing, and D. G. Robertson (1982). 'Interactive load forecasting and reserve allocation.' *CIGRE, paper 32–23, September 1982.* pp. 1–9.
2. R. M. Dunnett (1983). 'Economic allocation of reserve in a thermal power system with pumped storage and gas turbine plant.' *Third Int. Conf. Reliability of Power Supply Systems, London, 19–21 September, 1983.*
3. E. D. Farmer (1963). 'A method of prediction for non-stationary processes and its application to the problem of load estimation.' *Proc. 2nd IFAC Congr., Basle.*
4. R. E. Kalman (1960). 'A new approach to linear filtering and prediction problems.' *Trans. ASME*, Series D, **82,** 35–45.
5. E. D. Farmer, W. D. Laing, A. M. Adatia, A. B. Baker, and D. W. Bunn (1981). 'The development and implementation of an interactive demand validation and prediction facility,' *Seventh Power Systems Computation Conf., Lausanne, 12–17 July, 1981.* pp. 588–598.
6. A. B. Baker (1977). 'Forecasting for the operational requirements of the CEGB national grid system.' *Proc. Public Utilities Forecasting Conf., Bowness-on-Windermere, March, 1977.*

APPENDIX GENERATING SPOT VALUES FROM AVERAGES

If $\{D_i, 1 \leq i \leq N\}$ is a sequence of demand averages over N intervals, and if each average covers M sub-intervals, then spot values are required for the $M \times N$ sub-intervals such that, if the interval averages are recalculated, the original sequence is recovered. The total period covered by the sequence is represented by piecewise linear functions from which the spot values are interpolated, the slope and intersection of each line being constrained by the

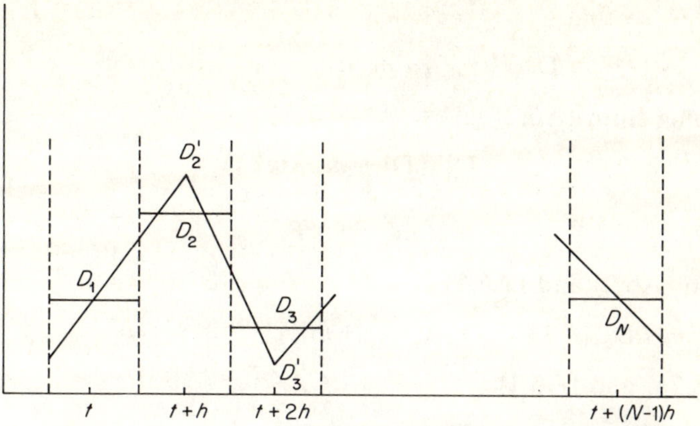

FIGURE A6.1 Spot values interpolated from averages

interval averages. If the lines intersect at the midpoint of each interval, then the points of intersection determine the parameters of the line. It is the values at these intersection points that have to be calculated, although the values at the boundary intervals are set to the original averages to remove the two degrees of freedom which exist. The problem is shown in Figure A6.1.

At the ith intermediate interval,

$$m_{i-1}a_{i-1}+c_{i-1}=m_i a_{i-1}+c_i \qquad 1<i<N \tag{A6.1}$$

where

$$a_i=t+ih$$

and

t is the midpoint time of the first interval
h is the interval length
m_i, c_i are the slopes and intercepts, respectively, of the lines.

The interval average value constraint gives, for the intermediate intervals,

$$D_i=\{(a_{i-1}-h/4)m_{i-1}+c_{i-1}+(a_{i-1}+h/4)m_i+c_i\}/2 \tag{A6.2}$$

for $1<i<N$.

At the two boundary intervals,

$$D_1=m_1a_0+c_1 \tag{A6.3a}$$

$$D_N=m_{N-1}a_{N-1}+c_{N-1}. \tag{A6.3b}$$

The new internal interval midpoint values are given by

$$D'_i=m_{i-1}a_{i-1}+c_{i-1}=m_ia_{i-1}+c_i$$

for $1<i<N$, so that

$$D'_i=\{(m_{i-1}+m_i)a_{i-1}+c_{i-1}+c_i\}/2. \qquad \text{(A6.4)}$$

Substituing from (A6.2) gives

$$D'_i=D_i-(m'_i-m'_{i-1}) \qquad \text{(A6.5)}$$

where

$$m'_i=hm_i/8.$$

Now, from (A6.2) and (A6.1)

$$D_{i+1}-D_i=6m'_i+m'_{i+1}+m'_{i-1} \qquad 2\leqslant i\leqslant N-2. \qquad \text{(A6.6a)}$$

From (A6.3a) and (A6.1)

$$D_2-D_1=7m'_1+m'_2. \qquad \text{(A6.6b)}$$

From (A6.3b) and (A6.1)

$$D_N-D_{N-1}=7m'_{N-1}+m'_{N-2}. \qquad \text{(A6.6c)}$$

Ordering (A6.6), subtracting successive pairs, and eliminating m' using (A6.5) leads to

$$D'_3+6D'_2=8D_2-D_1 \qquad \text{(A6.7a)}$$

$$D'_{i+1}+6D'_i+D'_{i-1}=8D_i \qquad i=3,\ldots,N-2 \qquad \text{(A6.7b)}$$

$$6D'_{N-1}+D'_{N-2}=8D_{N-1}-D_N \qquad \text{(A6.7c)}$$

which are $N-2$ equations in the $N-2$ unknowns D'_i $(1<i<N)$.

The system of equations (A6.7) has a symmetrical tri-diagonal structure which can be solved efficiently using special techniques. The spot values are calculated by interpolating along the lines joining the modified mid-interval demands, D'_i, and the two boundary values, D_1 and D_N.

Comparative Models for Electrical Load Forecasting
Edited by D. W. Bunn and E. D. Farmer

CHAPTER 7

24-hour Electric Utility Load Forecasting

Alan M. Schneider, Tadashi Takenawa, and Dean A. Schiffman

1 INTRODUCTION

In recent years, load forecasting for electric power systems has become increasingly important. Load forecasting is necessary for economic generation of power, economic allocation between plants (unit commitment scheduling), maintenance scheduling, and for system security such as peak load shaving by power interchange with interconnected utilities.

The objective of the study described in this chapter was to develop a model which is suitable for on-line hourly load forecasting over a 1 to 5-day period. The proposed model represents an advance in the state-of-the-art in that the implementation of the model in a digital computer is much easier than those described in References 1–3, and it is capable of utilizing Weather Service temperature forecasts (daily high and low) for hourly load forecasting.

2 BACKGROUND

The model has been implemented in a digital computer program for forecasting next-day hourly load profiles for the electric utility serving the San Diego area, the San Diego Gas and Electric Company (SDG & E).

Key geographic and demographic features of San Diego are described as an aid in understanding the real-world situation for which this model was developed. The SDG & E service territory is roughly contiguous with the boundaries of the County of San Diego, a region of approximately 1.7 million people. The load is about evenly divided between residential, commercial (including apartments), and industrial users. The winter hourly peak of about 1500 megawatts (MW) occurs in the evening, typically around 6 o'clock in December. The summer peak of about 1500 MW is due to air conditioning and occurs typically at about 2 p.m. in July and August. The region has been experiencing about 7.1% annual growth rate in electrical energy use since May, 1974.[4]

San Diego County is roughly 60 miles square. Two-thirds of the population is located in and around San Diego in a 20-mile square. Climate varies considerably over this region. The coastal strip, perhaps a mile wide, bordering the Pacific, has a moderate climate, with typically no more than a 15 °F temperature differential between night and day and with frequent fog. Inland of this are the mesas, 400–700 ft high, which are warm and sunny throughout most of the year, with a day–night temperature differential of 25–30 °F, extending 20 miles in from the ocean. In the hottest part of a summer day, there is very nearly a one-degree-per-mile temperature rise as one travels inland. It is not uncommon to find cloudy, overcast, or foggy conditions in coastal La Jolla or Point Loma while the sun is shining brightly a mile to the east. The variability of climate from place to place is termed 'micro-climate'. Most of the year's 10 in of rainfall occurs in December to April. A small part of the population lives in the mountains which range up to 4500 ft and run parallel to the coast, 40–50 miles inland. East of the mountains, there is the desert, with yet another climate. Agriculture in coastal, mesa, mountain, and desert communities accounts for part of SDG & E's load. Several times a year, the usual wind-pattern of mild, humid ocean-to-land daytime breezes will be reversed and, for a period of three to five days, the region experiences hot, dry, desert-to-ocean daytime breezes. This condition, called a 'santana', causes the temperature to be unseasonably warm; the relative humidity can drop from its usual 60–70% to 10–15%. Skies will be cloudless and often smoggy.

For many years, an experienced person at SDG & E has made a daily forecast of the next day 24-hourly load profile. The objective of the modelling which is the subject of this chapter is to automate this process.

3 OVERVIEW OF THE LOAD-FORECASTING MODEL

The database consists of the past values of total system hourly load in MW and the hourly temperature readings at a point near the geographic and population centre. The database is used to establish the model parameters, constant for any given forecast. The model consists, mathematically, of four components: (a) the 'basic' load, (b) the temperature-deviation component, (c) the noise component ('residual'), and (d) the bias correction. The basic load is the largest component, accounting for 90–95% of the total load. The temperature-deviation component accounts for 5–10% of the load. It reflects the change in load due to today's weather being different from that expected. The residual typically accounts for no more than 2–3% of the total. The fourth component, the bias correction, is normally zero. It comes into use only under special and unusual conditions—major city-wide events, holidays, unusual weather patterns such as rain or santanas, etc.

The model is also broken down into weekday (Monday 0100 to Friday

2400) and weekend (Saturday 0100 to Sunday 2400) models, which are similar in structure, but with different model parameters. Results using the two models proved to be superior to those obtained with a single weekly model.[5]

Our statistical (regression) analysis revealed that temperature is the leading factor among all weather variables and that its effect is pseudolinearly related to load. Windspeed and relative humidity usually affect the load much less. However, relative humidity can become important when the temperature is high in summer. Since it is difficult to forecast hourly windspeed and/or relative humidity, and these play a much smaller role than temperature, temperature is the only weather variable used in this model.

The temperature model is a 24-point model fit to the daily temperature cycle. Fitting this model over the model-fitting interval (typically the past three weeks) generates what is called the 'basic' temperature. The basic temperature $T_B(t)$ is the average or expected temperature, hour by hour, in the 3-week interval. The temperature-deviation component of the load depends on the difference, $\Delta T(t)$, between the basic temperature and the actual hourly temperature $T(t)$. Thus, if the actual temperature over a 3-week interval followed the basic temperature curve exactly, the temperature deviation $\Delta T(t)$ would be zero, as would the temperature-deviation component of the load. Assuming there were no holidays or special events in this period, the load in such a case would be modelled by the basic load. We would say, since the actual temperature equalled the expected, that the load pattern would be 'standard', and hence modellable by the basic load. When the actual temperature deviates from the expected, i.e. the basic, then and only then is there a correction made in the load. This correction is the temperature-deviation component.

Greater forecasting accuracy could probably be obtained if individual models were built for separate geographic regions. However, SDG & E does not monitor the load in such a way as to provide the needed data.

Before settling on the form of the model described here, two other types of forecasting models were tested: a Box–Jenkins type, similar to the one used in Reference 6, and a multiple-regression type, similar to those in References 7–12. The former was discarded because the forecast error tends to diverge as the period over which the forecast is made increases. The second was discarded because of the difficulty of generating input variables (such as future weather information) on an hourly basis.

4 DETAILED DESCRIPTION OF THE MODEL

The load model is

$$L(t) = L_B(t) + L_T(t) + U(t) + L_c(t) \tag{7.1}$$

where t = time of day in hours ($t = 1, 2, 3, \ldots, 24$), $L(t)$ = total load at hour t in MW, $L_B(t)$ = basic load at time t, $L_T(t)$ = temperature-deviation component of load at t, $U(t)$ = residual load at t, $L_c(t)$ = bias correction, normally zero, but otherwise accounting for unusual conditions. The load model is used in two modes, namely model-fitting and forecasting. In model-fitting, the model-fitting parameters are determined by using actual past data for load, $L(t)$, and temperature, $T(t)$. That is, the parameters which appear in the terms of the right-hand side (r.h.s.) of equation (7.1) are adjusted to get a good fit to the known load data, $L(t)$, inserted on the left-hand side (l.h.s.). In forecasting, on the other hand, future values of time t and forecasts of hourly temperature $T(t)$ are inserted on the r.h.s. of equation (7.1) to compute forecasts $L(t)$ of the load. Thus, $L(t)$ represents actual data when t is in the past, and forecast data for t in the future. The other terms in equation 7.1 have a similar dual interpretation.

Let us consider the components of the load one by one. The basic load is modelled by

$$L_B(t) = a_0 + \sum_{k=1}^{m} (a_k \cos \omega_k t + b_k \sin \omega_k t) \tag{7.2}$$

where $m = 9$ appears adequate to fit the data in the San Diego region. The ω_k are the fundamental and selected higher harmonics for the 5-day weekday period and 2-day weekend period as listed in Table 7.1. The a_k and b_k are $2m+1$ model-fitting parameters.

Figure 7.1(b) shows actual load and basic load over one day (along with other information to be discussed later). One does not expect good agree-

TABLE 7.1 Harmonics and corresponding frequencies in the basic load: T = fundamental period (hr), f_k = kth harmonic, ω_k = kth frequency (radian per hour) = f_k/T

	Weekday model $T = 120$	Weekend model $T = 48$
k	f_k	f_k
1	1	1
2	2	2
3	5	3
4	10	4
5	15	6
6	20	8
7	25	10
8	30	12
9	40	16

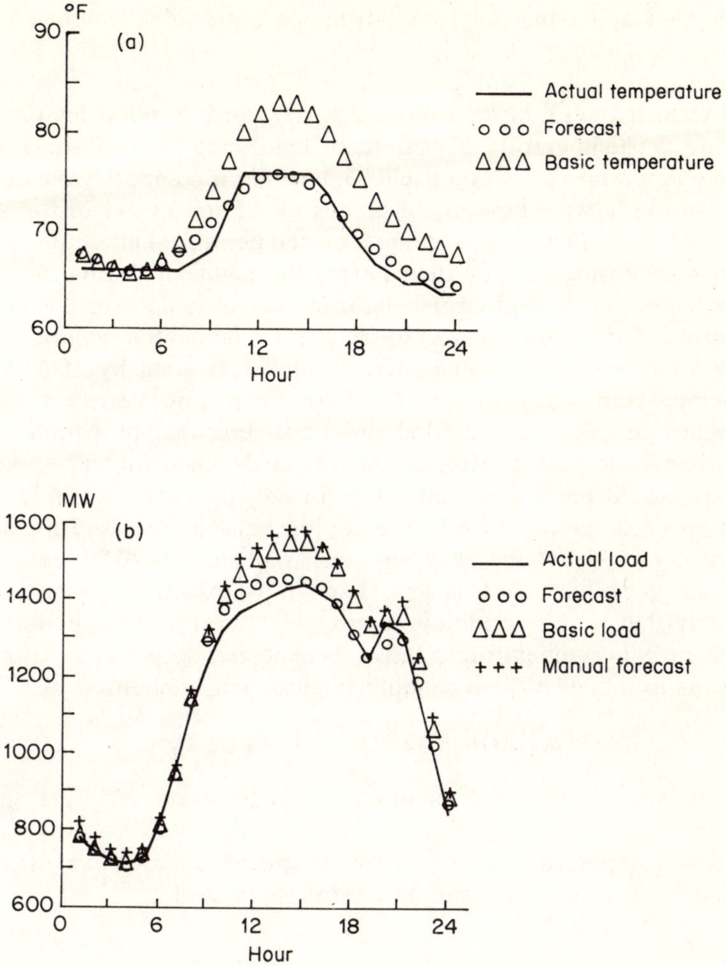

FIGURE 7.1 Forecasting of (a) temperature and (b) load on 9 August 1977

ment between basic and actual, because the temperature deviation component is missing.

The second term on the r.h.s. of equation (7.1), the temperature deviation component, depends on the following quantities: $T(t)$ = actual or forecast temperature at time t, $T_B(t)$ = basic temperature at t, $\Delta T(t)$ = temperature deviation at time t, relative to basic, $\Delta T_E(t)$ = effective temperature deviation.

The temperature deviation at time t is defined to be the difference between basic and actual (or forecast) temperature, viz.,

$$\Delta T = T(t) - T_B(t). \tag{7.3}$$

The effective temperature deviation, $\Delta T_E(t)$, which is the primary input in calculating the temperature deviations of load, may be equal to or smaller than the temperature deviation itself. The former is computed as a nonlinear function of the latter. These modifications to ΔT are introduced to account for the observed fact that a degree of temperature difference causes a different load during the day than during the night, or during the summer than the winter, or, in a given season, if the actual temperature is hot, cool, or moderate. For example, it was found that in the heating season, 1 degree decrease of temperature would increase the peak load by 10 MW if the actual temperature was below 65 °F, whereas it had little effect if the actual temperature equalled or exceeded this level. Presumably, people will not call for changes to their heating or lighting loads when winter temperatures are moderate, even if different from those expected. Similarly, a 1 °F increase in peak temperature in the cooling season increased the daytime peak load by 15 MW if the peak temperature exceeded 70 °F, but does not change the peak load if T is less than 70 °F. Again, the explanation is presumably that no air conditioners are on at the lower temperatures and hence changes in temperature relative to expected do not change the load. This led us to model ΔT_E as a nonlinear function symbolized by

$$\Delta T_E(t) = F_s[\Delta T(t) \mid T(t), T_B(t), t]. \tag{7.4}$$

Further details on how ΔT is modified to generate ΔT_E are given in Appendix A.

The basic temperature itself is a two-frequency Fourier expansion fit to the actual temperature over the model-fitting interval

$$T_B(t) = \alpha_0 + \sum_{k=1}^{2} (\alpha_k \cos \lambda_k t + \beta_k \sin \lambda_k t) \tag{7.5}$$

where the α_k and β_k are the model-fitting parameters. The λ_k are the frequencies associated with a 24-hr fundamental and 12-hr harmonic. Figure 7.1(a) shows the actual temperature $T(t)$ and basic temperature $T_B(t)$ for one day, as well as a forecast, to be discussed later. The basic temperature profile is exactly the same for every day of the model-fitting interval. It represents the average daily temperature cycle for this period.

It was found in the initial analyses that the extra load induced by temperature deviations depended not only on the present value of ΔT_E, but on past values as well. Hence, the model for the temperature-deviation component of load includes the present plus several past hourly values of

ΔT_E. That is,

$$L_T(t) = \sum_{k=0}^{h} c_k \, \Delta T_E(t-k) \tag{7.6}$$

where h is typically 3 for San Diego data. The c_k add $(h+1)$ more model-fitting parameters; they have units of megawatts per degree F.

The bias correction term, $L_c(t)$, is a general-purpose term. We have experimented with this term in forecasting the load for the second and third hot day following an initial hot summer day.[13] The second and third day appear to demand an extra load whose peak value of 70 MW could perhaps be attributed to some psychological response of the customers—anticipating the worst after one severe day, they prepare for the next by resetting the air conditioner controls to call for more cooling. While experimentation has been carried out to validate the concept, the bias correction $L_c(t)$ is not included in the current working program.

The residual term $U(t)$ is assumed to be a zero-mean uncorrelated white noise, which accounts for the random difference between the stochastic load L and the deterministic model $L_B + L_T + L_c$.

In the various components of the load model, there are other parameters called 'control' parameters. These are parameters whose values we have chosen. They include such parameters as the temperatures in the nonlinear function F_s at which changes in the nature of the function occur. For example, in the cooling model, the control parameter T_1 (Appendix A) is the temperature at which saturation begins, and T_{max} is the temperature at which saturation is complete. While great effort has gone into establishing the current values, it may well be that experience with the program will reveal the need for further adjustment of these parameters.

5 MODEL-FITTING PROCEDURES

5.1 Basic temperature parameters

The first step in model-fitting is to obtain the parameters of the basic temperature model for the past data period. Using equation (7.5), in which typically three weeks of hourly temperature data are used, weighted least-squares (WLS) estimates are made of α_0, α_1, α_2, β_1, β_2, as described in Appendix B.

5.2 Effective temperature deviation

With the α_k and β_k determined, the basic temperature $T_B(t)$ is computed for each hour in 3-week period. By comparing it with the actual temperature

$T(t)$, the temperature deviation $\Delta T(t)$ is obtained. This is inserted into F_s, along with $T(t)$, $T_B(t)$, and t, and the effective temperature $\Delta T_E(t)$ is obtained for each hour.

5.3 Basic load parameters

Let us assume that the present time is in the middle of the week. Then we shall use the weekday load model for fitting and forecasting. There will be 3 weeks × 5 days = 15 days of data in the database for fitting the basic load. If there are any holidays in the database, say on a Tuesday, then the load and temperature from that day are removed and replaced with data from the first nonholiday Tuesday preceding the original data base. This is done (in the absence of the $L_c(t)$ term for holidays) to prevent the unusual load characteristics of the holiday from distorting the basic load pattern.*

We rewrite equation (7.1) using equations (7.2) and (7.6), omitting $L_c(t)$. Then

$$L(t) = a_0 + \sum_{k=1}^{m} (a_k \cos \omega_k t + b_k \sin \omega_k t) + \sum_{k=0}^{h} c_k \, \Delta T_E(t-k) + U(t). \quad (7.7)$$

The load $L(t)$ and the effective temperature deviation $\Delta T_E(t)$ are now known for every hour in the data base. These data are used to make least-square estimates of the a_k, b_k, and c_k in equation (7.7). Once the a_k and b_k are known, the basic load $L_B(t)$ can be computed for any time, past or future, from equation (7.2). The parameters of the weekend basic load model are derived in the same way, except that now the data base contains 2 days × 3 weeks = 6 days of data.

If forecasts are being made on Friday for the following Saturday, Sunday, and Monday, the procedure would be to run the program once to obtain the weekend model and ultimately the weekend forecasts. A second run of the program would be made to obtain the weekday model; it would be used in the preparation of the Monday forecast.

5.4 Check model adequacy

A procedure is described in Section 4 of Reference 14 based on F-statistics for checking to see whether there are sufficient terms included in the series expansions for the basic load and temperature deviation components, as given in equation (7.7).

* An alternative holiday replacement procedure would be to 'forecast' the holiday load, after the fact, using actual temperatures. This 'data' can be treated as a pseudo-measurement. The gain in this procedure is that it cuts the necessary data span to three weeks.

6 FORECASTING PROCEDURES

6.1 Hourly temperature forecasting

The load is heavily dependent on the temperature. By designing the basic load to include the normal weather effects for the current period, only the deviation in temperature from expected causes a nonzero value for the temperature-deviation component of load $L_t(t)$. For forecasting purposes, the deviation in temperature means the difference between the forecast hourly temperature, and basic temperature, the latter representing the expected temperature for the time period. The Weather Service issues forecasts normally at 4 a.m. and again at 4 p.m. The 4 a.m. bulletin forecasts these temperatures: (1) the high expected (at about 2 p.m.) that day, (2) the low expected (about 5 a.m.) on the following day, and (3) the next day's high. The 4 p.m. bulletin forecasts the next low, high, and low temperature respectively. Occasionally, when conditions warrant, the Weather Service will issue an extra bulletin which updates a previous forecast. Only the temperatures themselves are forecast; the time at which they are expected to occur is not given.

To generate an hourly forecast of the next-day temperature, we were guided by the following principles: (i) use temperature patterns of the immediate past, (ii) utilize the Weather Service forecasts to the fullest extent, (iii) allow for updating the temperature forecast at any time, say, if the Weather Service issues a correction bulletin and/or if the weather changes unexpectedly, (iv) calculation should be relatively simple, (v) the latest (actual) hourly temperature measurement should be the starting point for the forecast, (vi) we must be able to forecast several days in advance (say, for Saturday, Sunday, and Monday at 4 p.m. on Friday) even though Weather Service forecasts may not extend that far into the future. The means of obtaining the temperature forecast, $T_F(t)$, are described in Appendix C.

6.2 Load forecasting

Once the model-fitting parameters have been found, and the temperature forecast $T_F(t)$ is available for a selected forecasting period, the load forecast can be generated. The forecast and basic temperature are compared (equation (7.3)) and the deviation $\Delta T(t)$ obtained. The deviation is inserted into the nonlinear function F_s (equation (7.4)) to produce the effective temperature deviation $\Delta T_E(t)$. This is inserted into equation (7.6) to generate the temperature deviation component of load $L_T(t)$ for each hour in the forecasting interval. The basic load is generated with equation (7.2). In the future, when the bias correction $L_c(t)$ model has been more fully developed, it would be computed at this stage of the process. Then all components of the load are added to generate the total load forecast.

7 PROGRAMMING CONSIDERATIONS

Running time for WLS batch-processing to obtain the parameters of the weather and load model, and to forecast the weather and load for the rest of today and the next day, is 30 sec on a Burroughs 6700 computer.

8 RESULTS

The validation of the model proceeds by the following steps: (1) evaluation of the load forecasting model using actual temperature in place of forecast temperature, (2) evaluation of the load and temperature-forecasting models given true high and low daily temperatures as forecasts in place of Weather Service forecast high and low, and (3) evaluation of the load and temperature-forecasting models given Weather Service forecasts. Each of these steps includes the previous steps within it. An extensive study has been done for Step (1).[13,5] The results in this section pertain to Step (2). Step (3) will be done as the program is used on-line by SDG & E. While it is true that Weather Service forecasting errors will contribute to load forecasting error, it was not our purpose in this paper to develop methods of weather forecasting which are superior to those of the Weather Service.

A test program was carried out to evaluate the accuracy of the load and temperature forecasting with the current version of the program. Starting at 1600, forecasts of temperature and load were made for the next 32 hr, i.e. until midnight of the following day. Forecasts for the following day were extracted and graphed. This was done for six different weeks, two each in February, August, and October 1977. Figure 7.1 gives detailed results for one day. Temperature data appear in Figure 7.1(a). Actual and predicted loads are compared in Figure 7.1(b), which also shows the manual load forecast extracted from records of SDG & E. Figure 7.2 gives the results for 1 week in August. Reference 15 provides the complete results, which are similar to those given here. Figure 7.2(a) compares the forecast and actual temperatures. Figure 7.2(b) shows actual and forecast load. The root-mean-square (r.m.s.) temperature forecasting error for each day in the 6 weeks was, in general, less than 2 °F, with rare exceptions. Table 7.2 gives the r.m.s. hourly forecasting errors in MW, day-by-day, comparing the manual and automated methods over 3 weeks. From Table 7.2, we conclude that the model is working effectively, given perfect weather 'forecasts'. It should be noted that the manual method includes the effect of errors in Weather Service forecasts, whereas the automated method does not.

Plots of the actual load, as in Figure 7.2(b), reveal the presence of the

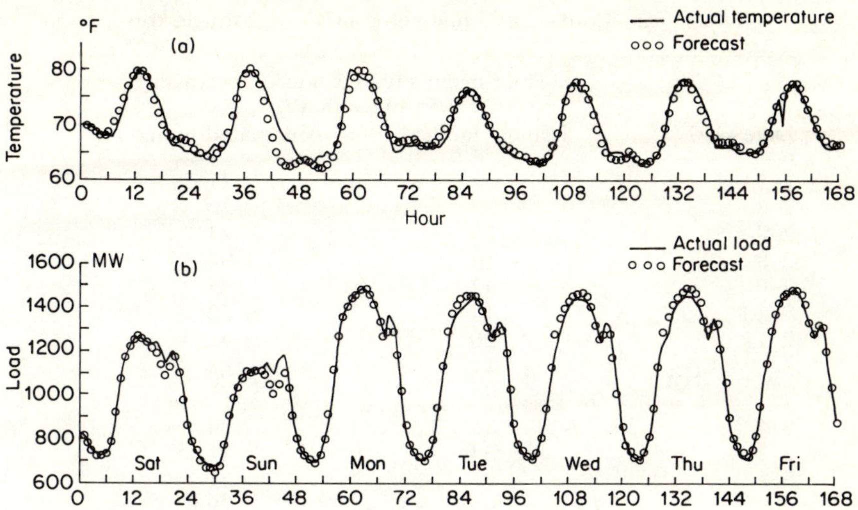

FIGURE 7.2 Forecasting of (a) temperature and (b) load for 6–12 August 1977

'evening peak', a narrow spike in the profile appearing at about $t = 20$ on each day. This is due to cooking the family meal electrically, turning on the household lights and television, and similar early evening activities. It would be possible to model this sharp spike in the load curve by including a large number of higher frequency terms in the Fourier series representing the basic load. However, it has been found to be more economical computationally to fit the evening peak by a Box–Jenkins[16] type of autoregressive model, which is, basically, a linear difference equation relating a sequence of random inputs $e(t)$ (samples of a zero-mean, independent, normally-distributed white noise process, the so-called 'shocks') to a random, but correlated, output sequence whose samples are denoted by $U(t)$. We therefore propose to subsume the evening peak in the term $U(t)$, which henceforth shall be called the 'evening peak' term, rather than the 'residual', in the following manner:

$$U(t) = \phi^{-1}(B)e(t) \tag{7.8}$$

where B = backward shift operator, such that $BU(t) = U(t-1)$; $\phi(B)$ = a general autoregressive operator (in essence, the transfer function of the difference equation); $e(t)$ = that component of load which the model fails to include. One of the advantages of this kind of model is the relatively few parameters typically needed to build an adequate model or transfer function. The parameters of the autoregressive operator are model-fitting parameters whose value is determined in the model-fitting process. Preliminary experiments have shown that two such parameters, one representing

TABLE 7.2 Root-mean-square of hourly forecasting errors

	Root-mean-squared hourly forecasting error (MW)	
Date (1977)	Manual method	Automated method
2–5	24	21
2–6	33	30
2–7	26	24
2–8	19	30
2–9	42	25
2–10	14	23
2–11	36	31
Over 7 days	29	26
8–6	35	24
8–7	34	41
8–8	84	21
8–9	95	29
8–10	32	25
8–11	24	33
8–12	35	30
Over 7 days	55	29
10–15	39	23
10–16	38	20
10–17	28	20
10–18	38	24
10–19	43	23
10–20	29	25
10–21	30	26
Over 7 days	36	23

the hourly fluctations and one representing a 24-hr periodicity, give good results. The estimation procedures of Appendix B must be modified slightly to account for the fact that $U(t)$ is not, as previously assumed, uncorrelated white noise. The modified procedure is described in Section 3 of Reference 14. The 'shocks' $e(t)$ are the 'residual' (i.e. noise) of this newly expanded model; a good model reduces the residual to a minimum in the mean-square sense.

A benefit of automation is the ease and speed with which forecasts can be made more often; in particular, revised forecasts can be made quickly if needed. Thus, for example, a forecast for today made at 7 a.m. could be updated again at 11 a.m. by de-weighting past data if the weather pattern took an unexpected turn and/or a Weather Service correction bulletin was available. Also, the ease of making 1-day forecasts on Saturday and Sunday could obviate the present need to make a 3-day forecast on Friday.

9 FURTHER CONSIDERATIONS

After having developed the program, and trying it out at SDG & E, we have identified areas in which improvements could be made. These are as follows.

1 Winter–summer transitions

In running the program, the 'effective temperature deviation' component of the model (see pp. 103–4) is placed in either a summer or winter mode. At certain times in the year (typically in the spring or autumn), it was not clear which mode would best apply. Therefore some means of transferring smoothly from one mode to the other would be desirable.

2 Replacing load data for unusual days

Implicitly, it is assumed that the 'basic load' component of the model is a stationary time series during the days over which the model is estimated. The parameters underlying this time series are assumed to depend only on whether the 'weekdays' or 'weekend' model is being considered.

Often, however, the load data collected correspond to unusual days such as holidays. (Consider, for instance, that Christmas Day and New Year's Day may both fall on, say, a Tuesday, which means that two of the three Tuesdays in the data base are atypical.) To remedy this situation, data from previous days of the same type (i.e. a Tuesday for a Tuesday, a Wednesday for a Wednesday, etc.) were substituted into the data-base along with corresponding data on temperature. Although the resulting hourly time series was not sequential, this method proved to be a useful approximation in the process of estimating the weekday or weekend models.

3 Incorporating improved weather data and forecasts

It is known that the afternoon weather pattern in San Diego is strongly correlated with the time of onset of the on-shore breeze, which occurs typically between noon and 3 p.m. If the actual time were noted on each day, the weather pattern for the remainder of that afternoon could be predicted more accurately. A means of incorporating this single piece of (deterministic) data would improve at least a part of the day's weather and load forecast.

Correlation studies should be undertaken between the load and other weather variables, such as cloud cover, humidity, and windspeed and direction. Since the Weather Service now forecasts the probability of rainfall, that information could be used in the forecasting process.

4 *Study user classes*

Our first study considered all users together. It would probably be useful to study the use patterns of residential, commercial, and industrial users separately. Then a theory similar to the one already developed could be applied to each user class. To implement this, however, requires that the actual hourly load data for each user class be available. Is it? Could it be?

5 *Separate geographical regions*

Because of the rather wide variation in weather from one part of San Diego to the next, it may be desirable to build separate models for separate regions, say, the coastal strip, the inland valleys, the mountains, and the deserts. As in the preceding paragraph, implementation of such a scheme necessitates having the hourly load data for the separate regions. We note that separate weather forecasts are now available for these geographical groupings.

6 *The psychological effect of a forecast of heavy clouds*

We noted in the first study the psychological effect of several hot days in a row, in which users would begin to anticipate heat, and call for more air conditioning (cooling) on later days, for equal temperature levels. There may be a psychological effect in which excess electricity is used in anticipation of darkness. This could occur, for example, if the weather forecast appearing on TV and in the newspapers warned of unusually heavy cloud cover. The effect needs to be studied.

7 *A single model for weekdays and weekends*

We separated the model into one 5-day weekday model and one 2-day weekend model. The frequency terms are completely different between these two models. Yet time itself is continuous between, say, Friday night and Saturday morning. It may be possible to have a single model which, while accounting for the differences between Friday and Saturday, for example, nevertheless, moves smoothly from weekdays to weekends and back. (See point 1, Winter–summer transitions, for a similar problem.)

ACKNOWLEDGEMENTS

The load-forecasting program has been developed over the last two years under the auspices of the Energy Center at the University of California, San Diego (UCSD) under grants from SDG & E. We appreciate the support and cooperation of Mr James Holcombe, Manager, Electric Operations Department, Messrs Robert Newsome and William Cook of that Department, and Messrs John Coker and William Goddard of the Operation Center of SDG & E, who provided direction, past load data, and insights on the causes of variation of load. Ms Jan Fix and Mr Dan Randall of CDC provided access to the CDC computation facilities. Mr Walter Russell, Air Pollution Meteorologist of the San Diego Air Pollution Control District, helped with temperature data.

REFERENCES

1. F. D. Galiana, E. Handschin, and A. Fiechter (1974). *IEEE Trans. Auto. Contr.*, **AC-19,** 887.
2. A. Keyhani and H. Daniels (1976). *Conf. Record No. 76CH 1057-9-REG5, Control of Power Systems Conference and Exposition.* p. 132.
3. V. Panuska and J. P. Koutchouk (1975). *Proc. 6th IFAC Cong., Boston.*
4. A. Fox, (1979) *M.S. Thesis,* Department of Applied Mechanics and Engineering Sciences, University of California at San Diego, La Jolla, California, p. 24.
5. T. Takenawa and A. M. Schneider (1978). *Report No. 2,* Energy Center, University of California at San Diego, La Jolla, California.
6. M. Hagan and R. Klein (1978). *IEEE Trans. Syst. Man. Cybern.*, **SMC-8,** 711.
7. G. T. Heineman, D. A. Norman, and E. C. Plant (1966). *IEEE Trans.*, **PAS-85,** 1144.
8. S. L. Corpening, N. D. Reppen, and R. J. Ringle (1973). *IEEE Trans.*, **PAS-92,** 1966.
9. J. Davey, J. J. Saacks, G. W. Cunningham, and K. W. Priest (1973). *IEEE Trans.*, **PAS-92,** 971.
10. C. E. Asbury (1973). Paper presented at *IEEE PES Summer Meeting, Vancouver, B.C., Canada.*
11. K. N. Stanton and P. C. Gupta (1970). *IEEE Trans.*, **PAS-89,** 951.
12. R. P. Thompson (1976). *IEEE Trans.*, **PAS-95,** 385.
13. T. Takenawa and A. M. Schneider (1978). *Report No. 1,* Energy Center, University of California at San Diego, La Jolla, California.
14. T. Takenawa and A. M. Schneider (1978). *Report No. 3,* Energy Center. University of California at San Diego, La Jolla, California.
15. A. M. Schneider, T. Takenawa, and D. Schiffman (1978). *Report No. 4,* Energy Center, University of California at San Diego, La Jolla, California.
16. G. E. P. Box and G. M. Jenkins (1970). *Time Series Analysis, Forecasting and Control,* Holden Day, San Francisco. Chs 3, 7, and 9.
17. J. M. Mandel (1973). *Discrete Techniques of Parameter Estimation,* Marcel Dekker, Paris.
18. E. D. Farmer and M. J. Potton (1968). *Proc. IEEE,* **115,** 1549.
19. E. D. Farmer and M. J. Potton (1966). *Proc. 3rd IFAC Cong., London,* 21F. 1.
20. W. R. Christiaanse (1971). *IEEE Trans.*, **PAS-90,** 9000.
21. K. L. S. Sharma and A. K. Mahalanabis (1974). *Proc. IEEE,* **121,** 59.
22. J. Toyoda, M. Chen, and Y. Inoue (1970). *IEEE Trans.*, **PAS-89,** 1678.
23. K. L. S. Sharma and A. K. Mahalanabis (1972). Paper presented at *IEEE PES Summer Meeting, San Francisco, California.*
24. T. Takenawa (1979). *Ph.D. Thesis,* Department of Applied Mechanics and Engineering Sciences, University of California at San Diego, La Jolla, California. Part II, Section 3.

APPENDIX A EFFECTIVE TEMPERATURE DEVIATION

Since it has been found[13,5] that the temperature deviation $\Delta T(t)$ is non-linearly related to the additional load which it causes, it is modified to the effective temperature deviation $\Delta T_E(t)$ by the nonlinear function F_s. The nonlinear function consists of three components, f_1, f_2, and f_3 (Figure A7.1). We use temperature information from one location in the imaginary centre of energy consumption for the whole system. To smooth out the high frequency fluctuations, five readings taken 12 min apart (two before, two after, and one at the hour) are averaged. The function f_1 modifies the main part of ΔT to ΔT_E. Different f_1 functions are used for summer and winter. The summer f_1 is depicted in Figure A7.2 for $\Delta T \geqslant 0$. ΔT_E is equal to ΔT when the basic temperature T_B is over a certain limit T_c, typically 70 °F, which says that a 1-degree increase in temperature relative to expected is fully effective in calling for increased load (plane b-c-e). ΔT_E is less than ΔT when $T_B < T_c$ (plane b-c-d). On this plane, actual temperature is greater than expected, so an additional load is called for, but not as much additional load as when basic temperature exceeds T_c, reflecting the fact that some air conditioners will not be on at the lower temperatures. T_0 is an arbitrary origin (any value less than T_R) and T_R is a reference temperature (typically, 65 °F) to which the plane a-b-c is bent to construct the plane b-c-d. T_R also defines the flat surface a-b-d along which there is no extra load even though actual temperature exceeds basic temperature. A similar function below the $T-T_B$ plane in Figure A7.2 describes f_1 for $\Delta T<0$.

The function f_2 is the night-time modulation function. Figure A7.3 shows this function vs. time of day. At midnight, any given temperature deviation ΔT has been found, in our studies, to have a smaller effect on load than the same deviation during the day, presumably because the industrial and commercial sectors are largely shut down and not responding to temperature deviations. Function f_3 takes care of saturation in summer and winter, that is, once all air conditioners (or heaters) are operating fully, further increases

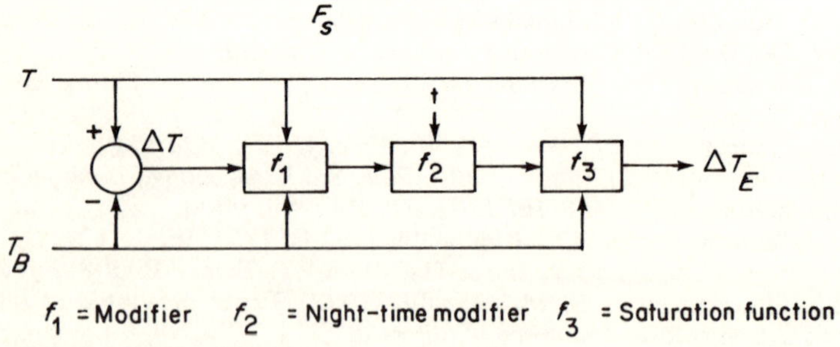

FIGURE A7.1 The effective temperature-deviation generator, F_s

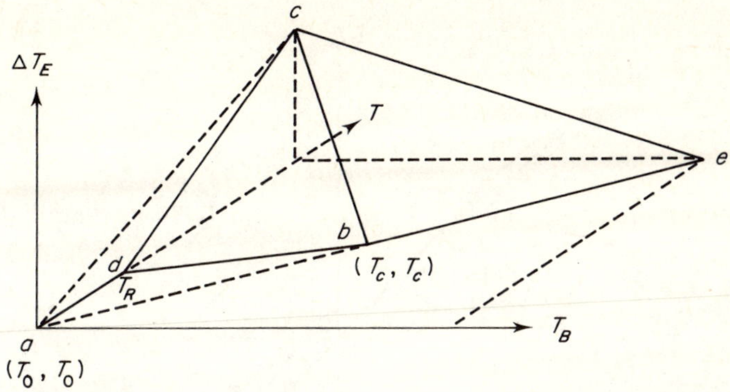

FIGURE A7.2 The nonlinear function f_1 for the case in which $\Delta T \geqslant 0$, for summer and for $f_2 = f_3 = 1$

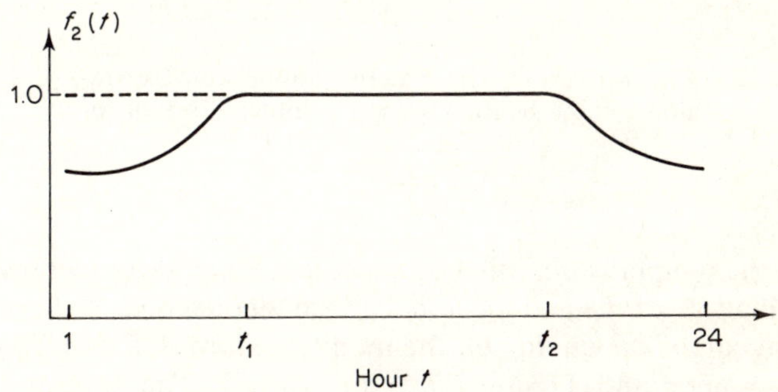

FIGURE A7.3 Night-time modifier, $f_2(t)$

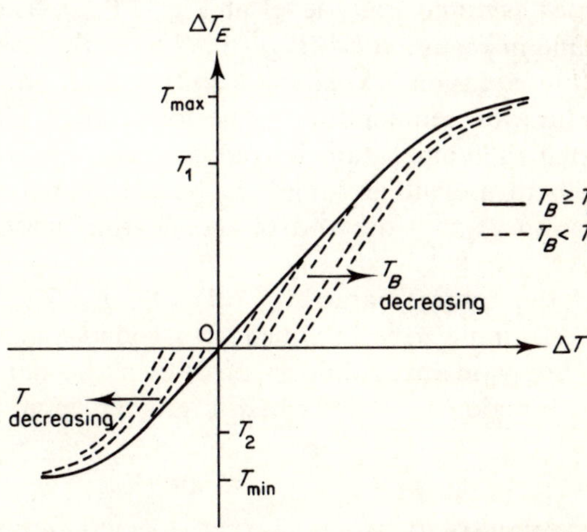

FIGURE A7.4 The effective temperature deviation for the cooling season (summer). (Drawn for the case where $f_2 = 1$)

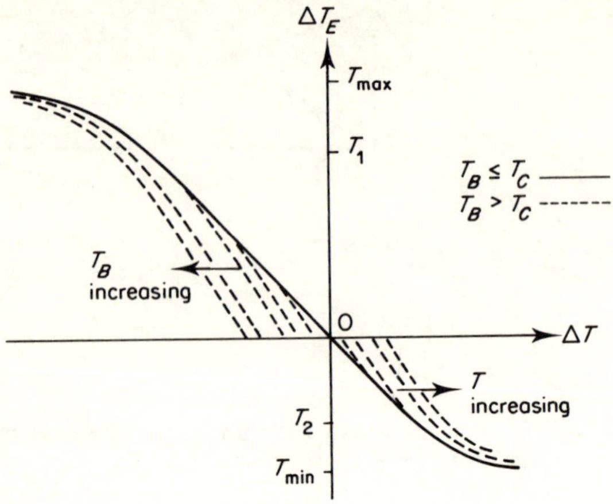

FIGURE A7.5 The effective temperature deviation for the heating season (winter). (Drawn for the case where $f_2 = 1$)

(decreases) in temperature will not cause corresponding increases in load. The combination of f_1 and f_3 also can be depicted as a family of two-dimensional curves depending on the modifying inputs T and T_B (see Figure A7.4 for summer and Figure A7.5 for winter). The saturation effect is represented by an exponential curve which begins at T_1 (T_2 on the negative side) and becomes asymptotically level at T_{max} (T_{min}). The origin 0 moves along the solid line in Figure A7.4 (Figure A7.5) as the average of the basic temperature (α_0 in equation 7.5) changes so that T_{max} and T_{min} correspond to the actual saturation temperatures. The low end of Figure A7.4 shows 'saturation' in that, if temperature is cold enough, no air conditioners are used. Lower temperature will no longer decrease the load. Similar reasoning holds for saturation at the low end of the heating season model, Figure A7.5.

The values of the control parameters, T_0, T_c, T_R, T_{max}, T_{min}, . . . , were selected after performing tests in which actual load and temperature data were used. The criterion was goodness of fit over the data span.

The nonlinear function F_s is described in greater detail in References 13 and 5.

APPENDIX B PARAMETER ESTIMATION

Assuming $L_c(t) = 0$ and $U(t)$ represents a zero-mean, uncorrelated white noise, the parameters used in fitting $L_B(t)$ and $L_T(t)$ to past data can be estimated by weighted least-square (WLS). Let

$$z_t = L(t)$$
$$x = [a_0, a_1, \ldots, a_m \vdots\, b_1, \ldots, b_m \vdots\, c_0, \ldots, c_h]^{\mathrm{T}}$$
$$h_t = [1, \cos(\omega_1 t), \ldots, \cos(\omega_m t) \vdots\, \sin(\omega_1 t), \ldots, \sin(\omega_m t) \vdots\, \Delta T_{\mathrm{E}}(t), \ldots, \Delta T_{\mathrm{E}}(t-h)]$$
$$u_t = U(t)$$

then

$$z_t = h_t x + u_t$$

or, in a vector form for k measurements,

$$Z = Hx + U \tag{B7.1}$$

where

$$Z = \begin{bmatrix} z_1 \\ z_2 \\ \vdots \\ z_t \\ \vdots \\ z_k \end{bmatrix} \qquad H = \begin{bmatrix} h_1 \\ h_2 \\ \vdots \\ h_t \\ \vdots \\ h_k \end{bmatrix} \qquad U = \begin{bmatrix} u_1 \\ u_2 \\ \vdots \\ u_t \\ \vdots \\ u_k \end{bmatrix}$$

The WLS estimate of x based on the past k measurements, denoted by $\hat{x}_k$, is given by

$$\hat{x}_k = (H^{\mathrm{T}}WH)^{-1}H^{\mathrm{T}}WZ \tag{B7.2}$$

where $W = \{w_{i,j}\}$ is a diagonal weighting matrix and $w_{t,t}$ is the weight given to the measurement z_t. Here, an exponential deweighting ($w_{t,t} = \beta^{k-t}$ where $0 < \beta \leqslant 1$, typically $0.99 \leqslant \beta \leqslant 1$) is used to emphasize more recent load patterns.

A recursive WLS algorithm for equation (B7.2) follows easily. Let

$$P_k = (H^{\mathrm{T}}WH)^{-1}. \tag{B7.3}$$

Clearly,

$$P_k^{-1} = P_{k-1}^{-1} + h_k^{\mathrm{T}} w_{k,k} h_k = P_{k-1}^{-1} + h_k^{\mathrm{T}} h_k. \tag{B7.4}$$

Using equations (B7.3) and (B7.4), equation (B7.2) can be rewritten as

$$\hat{x}_k = \hat{x}_{k-1} + P_k h_k^{\mathrm{T}}(z_k - h_k \hat{x}_{k-1}). \tag{B7.5}$$

Equation (B7.5) still requires the $M \times M$ matrix inversion $[M = \dim x = (2m+1) + (h+1)]$ to obtain P_k at each step k. However, applying the matrix inversion lemma[17] to equation (B7.4), P_k can be computed recursively as

$$P_k = \left(I - \frac{1}{v_k} P_{k-1} h_k^{\mathrm{T}} h_k\right) P_{k-1} \tag{B7.6}$$

where

$$v_k = h_k P_{k-1} h_k^T + 1. \tag{B7.7}$$

Substituting equation (B7.6) into equation (B7.5), $\hat{x}_k$ can be updated recursively without matrix inversions.

If the parameter vector is to be updated infrequently, either equation (B7.2) or equation (B7.5) can be used depending on the updating interval. The parameters in the basic temperature $T_B(t)$ can be computed similarly with the same weighting matrix W.

APPENDIX C 24-HOURLY TEMPERATURE FORECASTING

We desire to forecast 24-hourly temperatures, given the next day high and low forecasts provided by the Weather Service. To attack this problem, the process of fitting a Fourier series to the daily temperature has been studied. The power spectrum of the actual daily temperature has shown that daily and half-daily cycle terms are dominant, representing at least 90% of the power. As a result, the basic temperature has been modelled as a constant plus two frequencies, i.e.

$$T_B(t) = \alpha_0 + A_1 \cos \lambda_1(t + v_1) + A_2 \cos \lambda_2(t + v_2). \tag{C7.1}$$

Step 1 Find the tuned basic temperature

The tuned basic temperature, $T_B'(t)$, has exactly the same form but different parameter values, viz.

$$T_B'(t) = \alpha_0' + A_1' \cos \lambda_1(t + v_1') + A_2' \cos \lambda_2(t + v_2'). \tag{C7.2}$$

On the assumption that the most recent past is most indicative of the next-day temperature profile, another fit is made of the temperature data in which the past is more heavily deweighted ($\beta = 0.9$). The result is $T_B'(t)$. The times t_{min} and t_{max} at which this function reaches its minimum and maximum values must be found before proceeding.

*Step 2 Adjust the parameters of the tuned basic temperature to obtain the forecast temperature, $T_F(t)$, over one forecasting segment**

Assuming that the phase angles v_1' and v_2' in equation (C7.2) are unchanged for the next-day hourly forecasts, the tuned basic temperature is adjusted as follows to obtain the forecast temperature function:

$$\begin{aligned} T_F(t) &= d_0\alpha_0' + d_1 A_1' \cos \lambda_1(t + v_1') + d_2 A_2' \cos \lambda_2(t + v_2') \\ &= d_0\alpha_0' + d_1 A_1' \left(\cos \lambda_1(t + v_1') + \frac{d_2 A_2'}{d_1 A_1'} \cos \lambda_2(t + v_2') \right) \end{aligned} \tag{C7.3}$$

* A 'forecasting segment' is the time interval between successive high and low Weather Service forecasts.

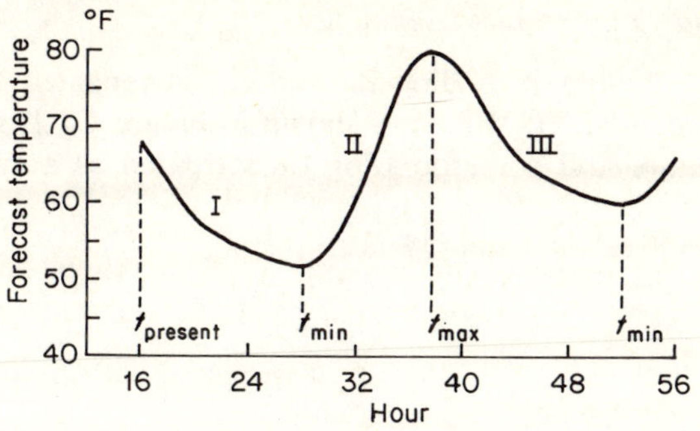

FIGURE C7.1 Hourly forecasting of temperature over three forecasting segments

where d_0, d_1, and d_2 are unknown adjustment coefficients to α_0', A_1', and A_2', respectively. These coefficients must be chosen so that the forecast temperature $T_F(t)$ in equation (C7.3) obtains a maximum and a minimum equal to the high and low Weather Service temperature forecasts for the segment. Then equation (C7.3) must satisfy the relation

$$T_F(t_i) = d_0\alpha_0' + d_1A_1'\left(\cos\lambda_1(t_i+v_1') + \frac{d_2A_2'}{d_1A_1'}\cos\lambda_2(t_i+v_2')\right) \qquad i = \max, \min \tag{C7.4}$$

where $T_F(t_{max})$ = high forecast and $T_F(t_{min})$ = low forecast, and

$$\left.\frac{dT_F}{dt}\right|_{t_i} = 0 = -d_1A_1'\left(\lambda_1\sin\lambda_1(t_i+v_1') + \frac{d_2A_2'}{d_1A_1'}\lambda_2\sin\lambda_2(t_i+v_2')\right)$$

or

$$\frac{\lambda_1\sin\lambda_1(t_i+v_1')}{\lambda_2\sin\lambda_2(t_i+v_2')} = -\frac{d_2A_2'}{d_1A_1'} \qquad i = \max, \min. \tag{C7.5}$$

Since no information is given for t_{max} and t_{min} for the next day by the Weather Service and, on the average, t_{max} and t_{min} fluctuate little day by day, it is reasonable to assume that

$$t_{max}, t_{min}\,|_{T_F} = t_{max}, t_{min}\,|_{T_B}. \tag{C7.6}$$

This result implies that $d_2/d_1 = 1$. Equation (C7.4) represents two equations in two unknowns, namely, d_0 and d_1. These equations are readily solved by evaluating at the end points of the forecasting segment, where $T_f(t)$ is available as either the present known value or a Weather Service forecast value. Finally, having determined all parameters on the r.h.s. of equation (C7.3), the forecast temperature function is available over one segment.

Step 3 Expand to all required segments

The procedure of Step 2, finding d_0 and d_1, is repeated for all required forecasting segments I, II, III, . . . as shown in Figure C7.1, with the present actual temperature and corresponding time treated as a 'Weather Service forecast.'

Note:

This chapter is based on a paper by the authors in *Energy*. It is reprinted with permission from *Energy*, **5,** 7, T. Takenawa, A. Schneider, and D. Schiffman, 'A computer program for 24-hour electric utility load forecasting', Copyright 1980, Pergamon Press Ltd.

Comparative Models for Electrical Load Forecasting
Edited by D. W. Bunn and E. D. Farmer

CHAPTER 8

Comparison of ARMA and Extended Wiener Filtering for Load Prediction at ENEL

U. Di Caprio, R. Genesio, S. Pozzi, and A. Vicino

INTRODUCTION

This paper specifically analyses the problem of on-line prediction (with a lead time from 1 min to 30 min) of the residual load obtained after the removal of a base load determined by using standard techniques.[1] This residual load usually exhibits nonstationary behaviour. For this reason, after the formulation of a current method based on stochastic ARMA models, Section 2 presents an alternative procedure derived from an extension of the Wiener filtering theory.[2–4] This extension allows one to deal with nonstationary components modelled by polynomial functions. In Section 3 a theoretical comparison is made between the two methods described in the preceding section and some conclusions about conditions under which each should be used are derived. Finally, Section 4 reports results referring both to real data (describing a load process measured in an area of Northern Italy) and to simulated data.

1 A PREDICTION METHOD BASED ON STOCHASTIC ARMA MODELS

Assume that the residual component of the load can be represented by the ARMA model

$$r(t)=\sum_{1}^{n}{}_{i}\,a_i r(t-i)+\sum_{1}^{m}{}_{i}\,b_i e(t-i)+b_0 e(t) \qquad (8.1)$$

where t is a discrete variable and $r(t)$ is the value of the residual load at time

This chapter is adapted from the same authors' paper 'Short-term Load Forecasting in Electric Power Systems: A Comparison of ARMA Models and Extended Wiener Filtering', *Journal of Forecasting*, **2**, 1983 (J. Wiley & Sons, Ltd), with permission.

t. The sequence $e(t-m), \ldots, e(t)$ is made up of independent random variables having the same distribution and characterized by zero mean and by unit variance. In (8.1) m and n are positive integers, a_i and b_i represent unknown coefficients which are allowed to vary with time, and finally b_0 is a constant with known value which essentially accounts for the variance of $r(t)$.

Equation (8.1) takes the following vector form (the symbol ^{T} means transpose):

$$r(t) = \mathbf{z}^{\mathrm{T}}(t)\mathbf{p}(t) + b_0 e(t) \tag{8.2}$$

where

$$\begin{aligned} \mathbf{z}^{\mathrm{T}}(t) &= |r(t-1), \ldots, r(t-n), e(t-1), \ldots, e(t-m)| \\ \mathbf{p}^{\mathrm{T}}(t) &= |a_1, \ldots, a_n, b_1, \ldots, b_m|. \end{aligned} \tag{8.3}$$

Assuming that $\mathbf{p}(t)$ satisfies the equation

$$\mathbf{p}(t+1) = \mathbf{p}(t) + \boldsymbol{\varepsilon}(t) \tag{8.4}$$

where $\boldsymbol{\varepsilon}(t)$ is an $(m+n)$ random vector having zero mean and covariance matrix equal to an assigned matrix $\mathbf{V}$, the minimum variance estimate $\hat{\mathbf{p}}(t)$ of $\mathbf{p}(t)$ obtainable from the sequence of observations $r(0), r(1), \ldots, r(t)$ turns out to be determined by the following recursive relationships:[5,6]

$$\begin{cases} \hat{\mathbf{p}}(t) = \hat{\mathbf{p}}(t-1) + \rho(t)\mathbf{P}(t-1)\mathbf{z}(t)[r(t) - \mathbf{z}^{\mathrm{T}}(t)\hat{\mathbf{p}}(t-1)] \\ \mathbf{P}(t) = \mathbf{P}(t-1) + \mathbf{V}(t-1) - \rho(t)\mathbf{P}(t-1)\mathbf{z}(t)\mathbf{z}^{\mathrm{T}}(t)\mathbf{P}(t-1) \\ \rho(t) = [b_0^2 + \mathbf{z}^{\mathrm{T}}(t)\mathbf{P}(t-1)\mathbf{z}(t)]^{-1}. \end{cases} \tag{8.5}$$

Under very general conditions this recursive procedure, starting from initial estimates of $\mathbf{p}(0)$ and $\mathbf{P}(0)$, is asymptotically stable. Also, it could be adapted with minor modifications so as to introduce an exponential decay to the contribution given by previous data. It is to be remarked that the vector $\mathbf{z}(t)$ must be estimated as $\hat{\mathbf{z}}(t)$, with the unknown $e(t-i)$ evaluated from (8.2) as

$$\hat{e}(t-i) = \frac{1}{b_0}[r(t-i) - \hat{\mathbf{z}}(t-i)\hat{\mathbf{p}}(t-i)]. \tag{8.6}$$

If the parameters a_i and b_i are constant, the optimal linear estimator $\hat{r}(t+\lambda)$ of the future value r at time $t+\lambda$ is determined by extending equation (8.1) up to time $t+\lambda$. The resulting forecasts, with the origin at t, can be written

$$\hat{r}(t+\lambda) = \mathbf{z}^{\mathrm{T}}(t+\lambda)\hat{\rho}(t) \tag{8.7}$$

where $\hat{e}(\tau) = 0$ for $\tau > t$. It can be shown that this form minimizes the variance of the prediction error.[7]

The preceding procedure assumes the *a priori* definition of the model structure, namely of the integers m and n in equation (8.1). This may be done through general criteria[8,9] which show that the optimal approximating structure results in a tradeoff between the number of parameters considered and the uncertainty due to their estimation errors; these in turn depend on the available information. In practice, if the process has a known structure with several unknown parameters, one could find it more convenient to represent it by a simplified model with few parameters ('principle of parsimony').

In the case of the load process, general criteria for structure determination cannot be satisfactorily used because of the series (possibly nonstationary and nonlinear). Therefore a trial and error procedure has simply been employed, which consists in determining the integers m and n so that the function

$$J = \sum_{0}^{N-1}{}_{\tau} [r(\tau+\lambda) - \hat{r}(\tau+\lambda)]^2 \tag{8.8}$$

where N is the number of available data, is minimized. The numerical value of J takes into account both the contribution coming from errors in the structure of model (8.1) and the uncertainty resulting from the estimation and prediction procedures. Since the latter satisfies optimality conditions, at least in the stationary case, J may be considered a significant index of the validity of the structure approximation.

It is worthwhile mentioning the possibility of using an alternative but equivalent procedure proposed in Wittenmark[10] which, for given integers m and n in (8.1), determines the structure of the prediction filter and directly derives recursive updating of the parameters of this filter rather than those of the model (8.1). The drawbacks of these procedures are that they may require a greater computational effort and that they do not provide information about the model of the load series.

2 THE ZADEH–RAGAZZINI PREDICTION METHOD

A preliminary analysis shows that the residual component $r(t)$ of the load time series may be represented by a continuous model of the following type:

$$r(t) = s(t) + f(t) \tag{8.9}$$

where $s(t)$ is a zero mean stationary process having correlation function

$$\varphi_s(\tau) = Ke^{-|\tau|/\Theta} + He^{-\omega_n\zeta|\tau|} \cos(\omega_0 |\tau| + \vartheta) \tag{8.10}$$

where $f(t)$ is a first-order polynomial with time-varying coefficients defined by

$$f(\tau) = \gamma_0(t) + \gamma_1(t)\tau \qquad t - T \leqslant \tau \leqslant t. \tag{8.11}$$

With regard to the generic time t, the positive constant T identifies the length of the past observation interval $(t-T, t)$ in which the polynomial $f(t)$ in (8.9) can be represented in the form expressed by (8.11).

The structure of the continuous model (8.9) leads one to resort to the Zadeh–Ragazzini prediction method which extends the classical Wiener theory to nonstationary processes characterized by polynomial nonstationary components.

2.1 Description of the method

The Zadeh–Ragazzini method[2] allows one to determine the mean square optimal linear filter for estimating the future value $\hat{r}(t+\lambda)$, given continuous time process observed in $(t-T, t)$ as

$$\hat{r}(t+\lambda)=\int_0^T h_\lambda(\tau)r(t-\tau)\,\mathrm{d}\tau \tag{8.12}$$

where λ is the lead time and $h_\lambda(\tau)$ is the impulse response of the optimal linear filter.

By assuming l as the order of the generic polynomial $f(t)$, $h_\lambda(\tau)$ can be determined in closed form solving the following system of $(l+2)$ equations:

$$\int_0^T h_\lambda(\tau)\varphi_s(t-\tau)\,\mathrm{d}\tau=(\mu_0+\mu_1 t+\ldots-\mu_l t^l)+\varphi_s(t+\lambda) \qquad 0\leqslant t\leqslant T \tag{8.13}$$

$$(-1)^i\lambda^i=\int_0^T \tau^i h_\lambda(\tau)\,\mathrm{d}\tau \qquad i=0, 1, \ldots, l. \tag{8.14}$$

(8.13) is generalization of the Wiener–Hopf equation. The constants $\mu_0, \mu_1, \ldots, \mu_l$ have the mathematical meaning of Lagrangian multipliers referring to the following constrained optimization problem: find the impulse response $h_\lambda(\tau)$ that minimizes the variance $\sigma_e^2(\lambda)$ of the prediction error while satisfying the condition

$$E[\hat{r}(t+\lambda)]=f(t+\lambda) \tag{8.15}$$

where E denotes the expected value. Observe that the coefficients of the polynomial need not be known: indeed $h_\lambda(\tau)$ turns out to be independent of the values of such coefficients.

It can be shown that $h_\lambda(t)$ has the following structure:[2]

$$\begin{aligned}h_\lambda(t)=[u(t)-u(t-T)]\Big[&A_0+A_1 t+\ldots+A_l t^l+B_1\,\mathrm{e}^{\beta_1 t}+B_2\,\mathrm{e}^{\beta_2 t}\\&+\ldots+B_{2m}\,\mathrm{e}^{\beta_{2m}t}+\frac{1}{2}R(p)\int_{-\infty}^{\infty}\frac{\phi_s(\omega^2)}{F(\omega^2)}R(-\mathrm{j}\omega)\,\mathrm{e}^{\mathrm{j}\omega(t+\lambda)}\,\mathrm{d}\omega\Big]\\&+C_1\,\delta(t)+\ldots+C_{n-m}\,\delta^{(n-m-1)}(t)+D_1\,\delta(t-T)+\ldots\\&+D_{n-m}\,\delta^{(n-m-1)}(t-T)\end{aligned} \tag{8.16}$$

where $u(t)$ and $u(t-T)$ are unit step functions indicating that $h_\lambda(t)$ is zero outside the interval $(0, T)$; $\delta^{(i)}(t)$ denotes the derivative of order i of the Dirac impulse function; $\phi_s(\omega^2)$ is the spectrum of $s(t)$; $R(j\omega)$ and $F(\omega^2)$ satisfy the relationship

$$\phi_s(\omega^2)=\frac{F(\omega^2)}{G(\omega^2)}=\frac{Q(j\omega)Q(-j\omega)}{R(j\omega)R(-j\omega)}=\left.\frac{Q(p)Q(-p)}{R(p)R(-p)}\right|_{p=j\omega} \tag{8.17}$$

where p is the derivative operator and $Q(p)$ and $R(p)$ are Hurwitz polynomials of order m and n respectively. The coefficients $\beta_1, \beta_2, \ldots, \beta_{2m}$ are the roots of the characteristic equation

$$F(-p^2)=0. \tag{8.18}$$

Finally, the $2n+l+1$ constants A_i, B_i, C_i, and D_i can be determined by solving a linear system of $2n+l+1$ equations formed by:

(a) $2n$ equations resulting from substitution of (8.16) in (8.13) and requiring that it is identically satisfied term by term
(b) $l+1$ equations resulting from substitution of (8.16) in (8.14).

Indeed, the preceding step (a) provides $2n+l+1$ equations, $2n$ of which are used in the way thus far explained, while the remaining $(l+1)$ allow one to compute the Lagrangian multipliers $\mu_0, \mu_1, \ldots, \mu_l$ as functions of $A_0, A_1, \ldots, A_l$ (see the example in the Appendix).

Though the multipliers μ_i do not appear in $h_\lambda(t)$, they keep a relevant practical meaning for their direct contribution to the minimum variance

$$\sigma^2_{e\,\min}(\lambda)=\sum_{i=0}^{l}(-1)^i\lambda^i\mu_i-\int_0^T h_\lambda(\tau)\varphi_s(\tau+\lambda)\,d\tau+\varphi_s(0). \tag{8.19}$$

(8.19) deserves the greatest attention, since it points out the explicit dependence of $\sigma^2_{e\,\min}$ upon the lead time λ and upon the length of observation interval T. Then it can be used as a design tool to *a priori* evaluate the maximum lead time for which the prediction is worth doing, or the minimum length of the observation interval necessary for $\sigma^2_{e\,\min}$ to be lower than a given threshold.

2.2 The asymptotic behaviour

It is known[2] that when $T\rightarrow\infty$ the impulse response $h_\lambda(t, T)$ as given by (8.16) becomes identical to that of the Wiener filter. Indeed, it can be checked that the coefficients of the exponentials and those of the Dirac functions vanish as $T\rightarrow\infty$. The only exception is represented by the term $C_1\,\delta(t)$ which weights the contribution given to $\hat{r}(t+\lambda)$ by the present value $r(t)$. Since the optimality of the Wiener filter is guaranteed only with regard to a purely stationary process, the asymptotic Zadeh–Ragazzini filter becomes unable to optimally predict the future of a process $r(t)$ containing a nonstationary component. On the other hand, this loss of optimality happens in the limit, namely for $T=\infty$, when the assumption on the polynomial trend degenerates.

If T is very large then the Zadeh–Ragazzini filter remains optimal (with regard to processes containing polynomial components) owing to the presence of the term $(A_0+A_1t+\ldots+A_lt^l)$ in $h_\lambda(t, T)$. This determines a substantial difference of structure with respect to the Wiener filter. More precisely, though the coefficients $A_0, A_1, \ldots, A_l$ become very small (see the example in the Appendix), the convolution of $(A_0+A_1t+\ldots+A_lt^l)$ with $r(t)$ may result in a significant contribution to $\hat{r}(t+\lambda)$, as soon as $r(t)$ contains a nonstationary polynomial component. Hence, while the nonpolynomial part of $h_\lambda(t, T)$ effectively approaches the impulse response of the Wiener filter as $T\rightarrow\infty$ (and thus can be practically replaced by the latter) it would be a mistake to neglect the polynomial part of $h_\lambda(t, T)$, however small the coefficients $A_0, A_1, \ldots, A_l$ turn out to be.

2.3 A recursive implementation

The structure and the parameters of the optimal filter are determined off-line, after having identified the degree of the polynomial that fits the nonstationary component and estimated the model and the parameter of the correlation function of the stationary component.

The impulse response $h_\lambda(\tau)$ results in a function which is equal to zero outside the interval $(0, T)$. In spite of this, the on-line computation of the predicted values $\hat{r}(t+\lambda)$ as a function of the running time can be generally made in recursive form. In fact, the integral (8.12) can be interpreted as the convolution of an extended impulse response, which represents the natural extrapolation of $h_\lambda(\tau)$ in $(0, \infty)$, with a signal equal to $r(\tau)$ in $(t-T, t)$ and zero elsewhere. When t varies, the above-mentioned extended impulse response does not vary. Hence, it can be shown that the discrete computation of $\hat{r}(t+1+\lambda)$, i.e. the predicted load from the observed data in $(t+1-T, t+1)$, is directly obtained as the output of a dynamic system characterized by the extended impulse response. Any step only requires for such systems a suitable updating of the state which provided the previous value $\hat{r}(t+\lambda)$. This updating is made according to the system dynamics and includes the disregarding of the contribution due to $r(t-T)$ and its replacement with the contribution given by the last data $r(t+1)$. So, only a number of data equal to the state dimension of the above dynamic system, usually much less than T, must be permanently stored in the prediction procedure.

3 COMPARISON OF THE TWO PREDICTION METHODS

The two prediction methods described in Sections 1 and 2 present different approach philosophies to the problem. The method based on ARMA models builds a discrete representation of the time series in recursive form (as the output of a suitable linear filter driven by stationary 'white noise'). Possible nonstationarities are taken into account by allowing a time variation of the parameters, which therefore must be successively updated. The mean square optimal prediction results from a simple time extrapolation of the ARMA model itself.

The Zadeh–Ragazzini approach describes the real process as a continuous signal resulting from the sum of a stationary stochastic process and non-stationary deterministic component expressed in a polynomial form. The parameters of the mean square optimal prediction filter are determined off-line by solving a generalized Wiener–Hopf equation.

The aim of this section is to theoretically compare the two methods in order to point out conditions for the use of one rather than the other. To this end a different form of the method of Section 1 based on ARMA models is here presented.

Introducing the forward shift operator q defined by

$$qr(\tau)=r(\tau+1) \tag{8.20}$$

equation (8.1) can be rewritten in the form

$$A(q^{-1})r(t)=b_0B(q^{-1})e(t) \tag{8.21}$$

where the polynomials A and B, assumed to be of the same degree ($m=n$), are equal to

$$A(q^{-1})=1-\sum_{1}^{n}{}_{i}\, a_iq^{-i} \tag{8.22}$$

$$B(q^{-1})=1+\frac{1}{b_0}\sum_{1}^{n}{}_{i}\, b_iq^{-i} \tag{8.23}$$

with all the zeros in the unit circle. Given the values of n, a_i, b_i, the minimum variance predictor at time $t+\lambda$ can be characterized by the following input–output relationship,[11] which corresponds to the classical Wiener filter:

$$B(q^{-1})\hat{r}(t+\lambda)=C(q^{-1})r(t) \tag{8.24}$$

where C is a polynomial having degree n and whose coefficients are defined by the identity

$$B(q^{-1})=A(q^{-1})D(q^{-1})+q^{-\lambda}C(q^{-1}) \tag{8.25}$$

More precisely, if

$$C(q^{-1}) = c_0 + c_1 q^{-1} + \ldots + c_n q^{-n}$$
$$D(q^{-1}) = 1 + d_1 q^{-1} + \ldots + d_{\lambda-1} q^{-\lambda+1} \qquad (8.26)$$

then the coefficients $c_0, c_1, \ldots, c_n$ and the coefficients $d_1, \ldots, d_{\lambda-1}$ can be determined by recursively solving equation (8.25).

The impulse response of the optimal predictor is then

$$h_\lambda(t) = \frac{C(q^{-1})}{B(q^{-1})}\delta'(t) \qquad (8.27)$$

where $\delta'(t)$ is the Kronecker delta function.

Assuming that the coefficients a_i and b_i in (8.1) are constant and known, the method described in Section 1 exactly implements the prementioned optimal predictor. In fact, the forecasting is obtained by extending equation (8.21) beyond the time t with application of equation (8.6) and (8.7). In practice, the predictor is implemented in a time-varying form since the parameters of (8.1) are updated at each step. These variations depend on $r(t)$ so that the final effect on the predictor turns out to be nonlinear and difficult to evaluate. Thus in the case where the parameters are constant but unknown, the prediction obtained by the method of Section 1 differs from that of filter (8.27) even if it generally yields the same asymptotical result.[12]

However, the more important source of error in forecasting may be derived from the fact that equation (8.1) does not provide an exact structural representation of the real process. This is the case in which the time series contains nonstationary components not included in the model (8.1) which can exhibit time-varying parameters. Such reasoning justifies a practical interest in evaluating the performance of the Zadeh–Ragazzini method which, on the contrary, prefigures the presence in the load model of polynomial components up to order l, without requiring *a priori* knowledge of their coefficients.

Since the Zadeh–Ragazzini filter minimizes the same error function taken into account by the ARMA model method, the limit value of $h_\lambda(t, T)$ as given by (8.16) for $T \rightarrow \infty$ is equal to the impulse response given by (8.27). Indeed, this relationship has already been proven[2] and takes into account the continuous formulation of the problem, as recalled in Section 2. The difference between $h_\lambda(t, T)$ and $h_\lambda(t, \infty)$ shown in Figure 8.1 measures the effect of the deterministic component of $r(t)$ which in turn defined by the two values of T and l.

Of course, the application of the Zadeh–Ragazzini filter suffers from errors analogous to those mentioned previously, with regard to the uncertainty in the knowledge of the real structure and of the parameters of the stochastic component. The correlation function $\varphi_s(\tau)$ is not exactly known,

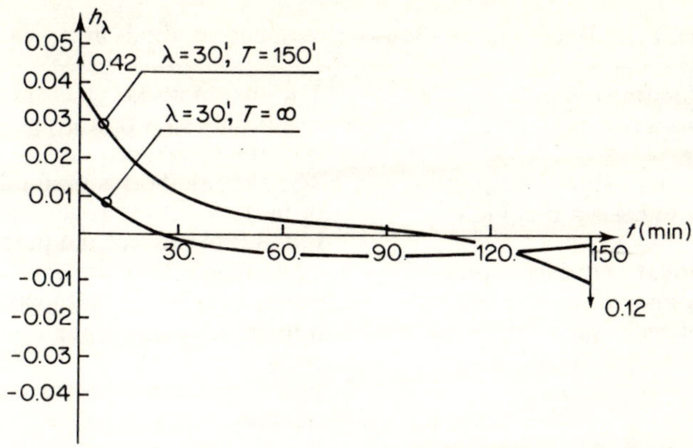

FIGURE 8.1 Diagrams of the impulse responses $h(t, T)$ and $h(T, \infty)$

but can only be estimated and the estimation error of $\varphi_s(\tau)$ results in a systematic error in forecasting because of the non-adaptive structure of the prediction filter. Another source of error is due to the fact that the nonstationary component of $r(t)$ is not exactly equal to a polynomial (with varying coefficients).

So, one is in the presence of two different methods which assume different models (a stationary stochastic model, and a nonstationary one) and both guarantee an optimal linear prediction, provided they are used in their nominal conditions. Really, both fail to satisfy these ideal conditions because of the errors in the model structure and of the uncertainty in the knowledge of the parameters. These errors are more likely minimized in the use of the first method because of its adaptivity. On the contrary, the second method has the advantage of being applicable to a much larger class of real processes.

Table 8.1 tries to assess the foreseeable behaviour of the two methods in each of four different typical cases.

The problem of the prediction of the electric load belongs to Case 4 of Table 8.1. Thus it seems difficult to state *a priori* which method will work better. The best that can be done is to analyse the numerical application to a real case, as shown in the following section, where some results from simulated data are also given.

4 RESULTS

In order to obtain an assessment of the effectiveness of the two methods, some experimental trials have been performed using real data. A time series of the electric load spanning five weeks has been recorded in an area of

TABLE 8.1 Behaviour of the two prediction methods in typical cases

Case 1: stationary stochastic process with known parameters	The two methods yield the same result provided that in the second $T=\infty$ is selected.
Case 2: stationary stochastic process with unknown parameters.	The first method is preferable because it can improve wrong initial estimates of the parameters.
Case 3: stationary stochastic process with known parameters, plus a deterministic component	The second method should work better because of its capability of directly accounting for deterministic components. The first method can in part follow the nonstationarities adapting its parameters.
Case 4: stationary stochastic process with unknown parameters, plus a deterministic component.	There is a superposition of the uncertainties pointed out in cases 2 and 3. Hence, the choice between the two methods depends on the relative weight of such uncertainties.

Northern Italy, with a maximum peak of 800 MW and a sampling interval of one minute. The residual component of the load referring to the fifth week was obtained using the base load computed with the previous four weeks. By means of a moving average with a window equal to 500 minutes, the stochastic component $s(t)$ of this residual load has been put in evidence. Its correlation function has then been computed upon a period spanning three days with a one-minute sampling interval (see Figure 8.2). Finally, a procedure based on the fitting of the correlation function has been used for parameter estimation.

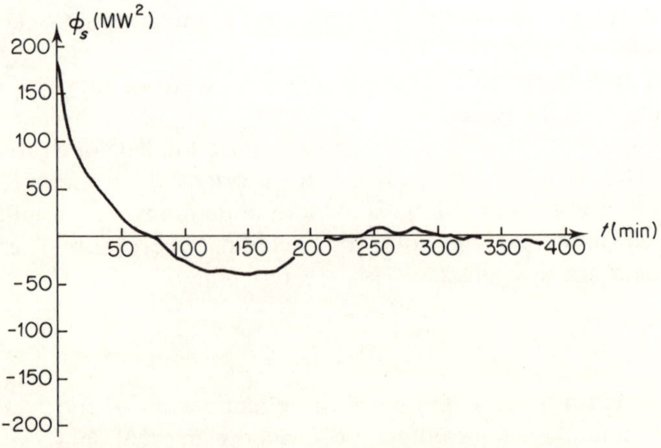

FIGURE 8.2 Estimate of correlation function $\phi_s(t)$

TABLE 8.2 Parameters of the correlation function $\varphi_s(\tau)$

Parameter	Estimated value	Standard deviation (%)
$K(\mathrm{MW}^2)$	94.1	22
$\Theta(\mathrm{min})$	7·4	33
$H(\mathrm{MW}^2)$	94.4	29
ζ	0.33	42
ω_n (rad/min)	0.02	10
ϑ (rad)	0.28	59

$\sigma_s = \sqrt{(K+H\cos\vartheta)} = 13.6$ MW

According to the model

$$\varphi_s(\tau) = Ke^{-|\tau|/\Theta} + H\,\mathrm{e}^{-\omega_n\zeta|\tau|}\cos(\omega_0|\tau| + \vartheta) \tag{8.28}$$

and letting

$$\omega_0 = \omega_n\sqrt{(1-\zeta^2)} \tag{8.29}$$

the results presented in Table 8.2 have been obtained.

The prementioned low-frequency component obtained from the moving average turns out to be well fitted by a first order polynomial, as shown by the numerical results reported in Table 8.3. They represent the fitting errors evaluated over 2750 samples and relative to different observation intervals T.

These errors are small with respect to the standard deviation σ_s of the stochastic component $s(t)$; that is, they should give little contribution to the forecasting error. Therefore, the nonstationary component has been modelled by a first-order polynomial ($l=1$) with time-varying coefficients. On the other hand, an increase of l, while slightly improving the approximation accuracy, gives rise to numerical problems in the solution of the Zadeh–Ragazzini filter.

In addition to the given experimental time series a realization of the stochastic process defined by the correlation (8.28) with the parameters of

TABLE 8.3 Fitting errors of the nonstationary term by a first order polynomial

T(min)	Average st. dev. (MW)	Min. st. dev. (MW)	Max. st. dev. (MW)
100	0.3	0.05	0.9
500	1.4	0.7	2.3
1000	2.1	1.4	3.3

TABLE 8.4 Ratio $\sigma_e/\hat{\sigma}_s$ for the Zadeh–Ragazzini filter

T(min)	$\lambda = 10$ min	$\lambda = 30$ min	T(min)	$\lambda = 10$ min	$\lambda = 30$ min
100	0.78	0.94	1000	0.83	0.98
300	0.83	0.87	3000	0.82	0.94
500	0.91	1.18	5000	0.80	0.93
700	0.83	1.03	10000	0.75	0.87

Table 8.2 has been simulated. The standard deviation of this time series was

$$\hat{\sigma}_s = 14.2\,\text{MW} \tag{8.30}$$

Referring again to the four different typical cases of Table 8.1, the results of forecasting obtained with real and simulated data are now presented.

4.1 Case 1. Stationary stochastic process with known parameters

From the simulated time series, the theoretical results for an ARMA model approach are those derived from the Wiener theory of prediction, i.e.

$$\begin{aligned} \sigma_e(\lambda = 10') &= 10.9\,\text{MW} \qquad \sigma_e/\hat{\sigma}_s = 0.77 \\ \sigma_e(\lambda = 30') &= 12.6\,\text{MW} \qquad \sigma_e/\hat{\sigma}_s = 0.89. \end{aligned} \tag{8.31}$$

The Zadeh–Ragazzini extension, as the observation interval $T \to \infty$, becomes a Wiener filter. This can be well recognized from Table 8.4 where the theoretical ratio of σ_e to $\hat{\sigma}_s$ has been reported for different lead times and observation intervals, with an assumed first order polynomial trend. The last row of the table can be considered corresponding to a Wiener filter, because of the large value of T. In fact, the reported results are very close to the theoretical values in (8.1).

4.2 Case 2. Stationary stochastic process with unknown parameters

Using the same data set of Case 1, an ARMA model of order (3, 3) has been used for process modelling, with initial values of the parameters different

TABLE 8.5 Parameter estimates for the correlation $\varphi_s(\tau)$

Parameter	Estimated value	Standard deviation (%)
$K(\text{MW}^2)$	106.4	23
Θ(min)	7.2	40
$H(\text{MW}^2)$	147.6	38
ζ	0.21	60
ω_n(rad/min)	0.02	12
ϑ(rad)	0.08	235

TABLE 8.6 Standard deviation of forecasting error

Prediction method	$\sigma_e(=10')$ (MW)	$\sigma_e(=30')$ (MW)
ARMA (3, 3)	11.4	13.5
Zadeh–Ragazzini, $T=100'$	12.1	17.1
Zadeh–Ragazzini, $T=1000'$	11.3	13.2

from the theoretical ones. After 3–4 hours (about 200 samples) the recursive algorithm yielded forecasting errors very similar to those of Case 1, as shown in Table 8.6 corresponding to the processing of about 5000 samples.

In the application of the Zadeh–Ragazzini method, a parameter estimation of the correlation function has been performed over the first 3000 values of the time series. The estimates are reported in Table 8.5. With these values of the parameters, different from the true values, an optimal predictor based on the Zadeh–Ragazzini theory has been impleemented. The final results of forecasting referring to about 5000 processed samples are reported in Table 8.6 for two different observation intervals T.

4.3 Case 3. Stationary stochastic process (with known parameters) plus a deterministic component

Using the previous data set (Cases 1 and 2) and summing the values obtained from the moving average to the simulated stationary realization, in order to reproduce real changes of electric load, a new time series has been derived. The two prediction methods have been tested on these data, which have a standard deviation equal to 19.2 MW, and a constant value of 44.0 MW was subtracted to allow the ARMA model to work with a zero mean process. The results obtained are shown in Table 8.7.

TABLE 8.7 Standard deviation of forecasting error

Prediction method	$\sigma_e(=10')$ (MW)	$\sigma_e(=30')$ (MW)
ARMA (3, 3)	11.9	15.3
Zadeh–Ragazzini, $T=100'$	12.7	17.9
Zadeh–Ragazzini, $T=1000'$	11.7	14.7

4.4 Case 4. Stationary stochastic process (with unknown parameters) plus a deterministic component

Up to now our concern has been with simulated data. On the contrary, Case 4 refers to real data of the electric load obtained after removal of the base load from the total load (see Figure 8.3). A preliminary analysis of the

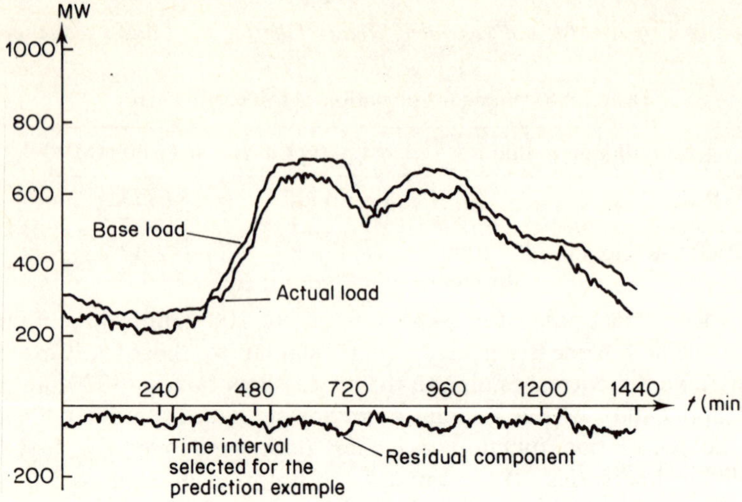

FIGURE 8.3 Time series of the electric load measured on Thursday 21 July 1977

TABLE 8.8 Standard deviation of forecasting error

Prediction method	$\sigma_e(=10')$ (MW)	$\sigma_e(=30')$ (MW)
ARMA (3, 3)	11.9	15.4
Zadeh–Ragazzini, $T = 100'$	13.4	19.6
Zadeh–Ragazzini, $T = 1000'$	13.1	16.5

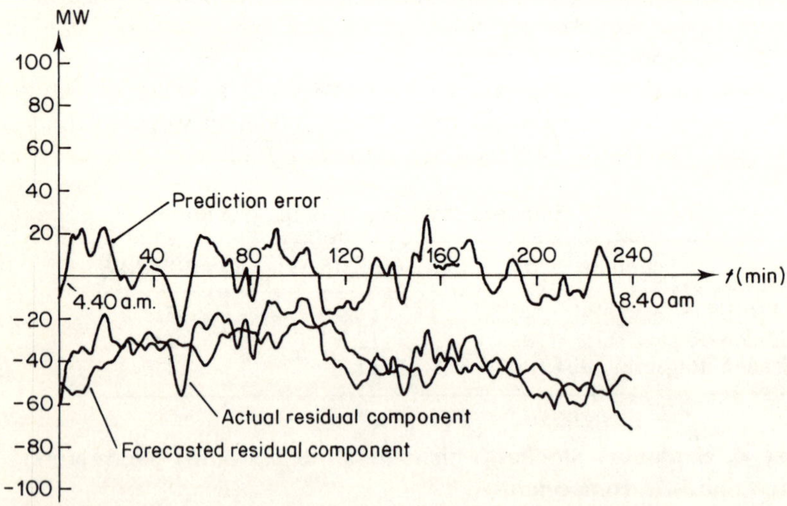

FIGURE 8.4 Example of prediction using the Zadeh–Ragazzini method: the lead time λ is equal to 10 min, while the observation interval is $T = 1000$ min

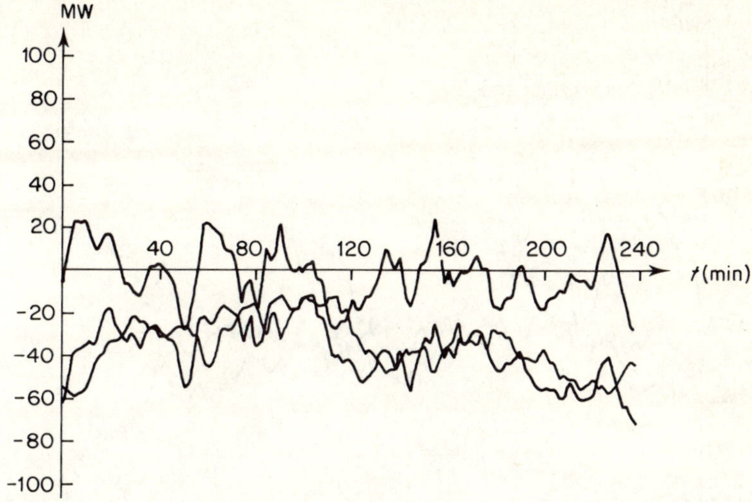

FIGURE 8.5 Example of prediction using an ARMA (3, 3) model: the lead time λ is equal to 10 min

data puts in evidence an average value of −44.0 MW and a standard deviation of 17.8 MW. The mean value is negative since the load of the fifth week is lower than the base load evaluated upon the four previous weeks: as the measurements have been performed in summer, load is particularly sensitive to the holiday slack.

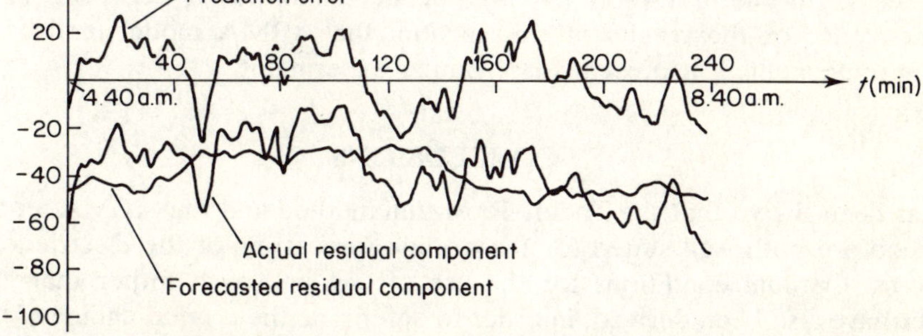

FIGURE 8.6 Example of prediction using the Zadeh–Ragazzini method: the lead time λ is equal to 30 min, while the observation interval is $T = 1000$ min

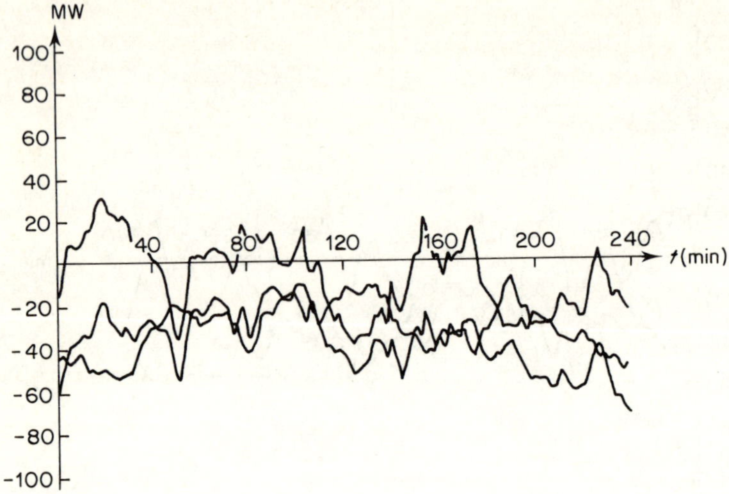

FIGURE 8.7 Example of prediction using an ARMA (3, 3) model: the lead time λ is equal to 30 min

The results obtained with the data spanning the fifth week are reported in Table 8.8 and examples of forecasting are shown in Figures 8.4–8.7.

Putting together the four numerical cases, one notices that the Zadeh–Ragazzini method works slightly better on the simulated data whereas the ARMA method is preferable on the real data. In this last condition, the use of an adaptive technique, as that employed by the procedure based on ARMA models, likely allows the prediction algorithm to better account for discrepancies between the assumed theoretical structure of the process and the real one.

Beyond the results described in this section, it is to be noted that the Zadeh–Ragazzini method has been implemented on a Varian 73 computer and tested on-line in a 4000 MW area of the Italian power network.[4] As a consequence of the results of this section, the ARMA model method is being implemented at present for a similar experiment.

5 CONCLUSIONS

It has been shown that the Zadeh–Ragazzini method and the ARMA model method are both well suited for the on-line forecasting of the electric load process. Optimal conditions for the use of one approach rather than the other have also been derived, in order to determine the *a priori* choice of the forecasting technique. These conditions depend upon the uncertainty of the model assumed for the time series description, with special regard to the behaviour of possible nonstationary components. The experimental results referring to the application of the two methods to a real load process are in close agreement with the preceding theoretical analysis.

REFERENCES

1. E. D. Farmer (1963). 'A method of prediction for non-stationary processes and its application to the problem of load estimation.' *Proc. 2nd IFAC Congress, Basle.*
2. L. A. Zadeh and J. R. Ragazzini (1950). 'An extension of Wiener's theory of prediction.' *J. Appl. Phys.*, **21,** iss. 7.
3. V. Marchese and S. Pozzi (1979). 'Short-term prediction of non-stationary load processes by the Zadeh–Ragazzini extension of the Wiener theory.' *Proc. 5th IFAC Symp. Identification, Darmstadt.*
4. R. Anelli, N. Maini, V. Marchese, and S. Pozzi, (1981). 'An on-line procedure for active load forecasting.' *Proc. PSCC, Lausanne.*
5. A. H. Jazwinsky (1970). *Stochastic Processes and Filtering Theory*, Academic Press, New York.
6. T. C. Hsia (1977). *System Identification*, Lexington Books.
7. G. E. P. Box and G. M. Jenkins (1970). *Times Series Analysis: Forecasting and Control*, Holden-Day, San Francisco.
8. H. Akaike (1974). 'A new look at the statistical model identification.' *IEEE Trans. on AC*, **AC-19,** iss. 5.
9. R. Genesio and M. Milanese (1979). 'Methods for the selection of approximating classes of models.' *Proc. 5th IFAC Symp. Identification, Darmstadt.*
10. B. Wittenmark (1974). 'A self-tuning predictor.' *IEEE Trans. on AC*, **AC-19,** iss. 6.
11. K. J. Åström (1970). *Introduction to Stochastic Control Theory*, Academic Press, New York.
12. K. J. Åström and B. Wittenmark (1973). 'On self tuning regulators.' *Automatica*, **9,** iss. 2.
13. M. S. Sadchev, R. Billington, and C. A. Peterson (1977). 'Representative bibliography on load forecasting.' *IEEE Trans. on PAS*, **PAS-96,** iss. 2.
14. E. D. Farmer and M. J. Potton (1968) 'Development of on-line prediction techniques with results from trials in the soutwest region of the CEGB.' *Proc. IEE*, **115,** iss. 4.
15. R. Anelli, U. Di Caprio, V. Marchese, and S. Pozzi (1978). 'Short term prediction of stationary load processes with a correlation function finite sum of exponentials.' *Proc. PSCC, Darmstadt.*
16. J. Toyoda, M. Chen, and Y. Inoue (1970). 'An application of state estimator to short-term load forecasting.' *IEEE Trans. on PAS*, **PAS-89,** iss. 5.
17. K. Srinivasan and R. Pronovost (1975). 'Short-term load forecasting using multiple correlation models.' *IEEE Trans. on PAS*, **PAS-94,** iss. 5.
18. W. R. Christiaanse (1971). 'Short term load forecasting using general exponential smoothing.' *IEEE Trans. on PAS*, **PAS-90,** iss. 2.
19. K. L. S. Sharma and A. K. Mahalanabis (1974). 'Recursive short-term load forecasting algorithm.' *Proc. IEE*, **121,** iss. 1.
20. S. Vemuri, E. F. Hill, and Balasubramanian (1973). 'Load forecasting using stochastic models.' *Proc. 8th PICA Conference, Minneapolis.*
21. A. Keyhani, R. E. Nasburg, A. H. E. El Abiad, and R. L. Kashyap (1975).

'Stochastic load modelling for on-line applications: part I and part II.' *Proc. Eng. Found. Conference, Lafayette.*
22. U. Panuska (1975). 'Electrical power system load modelling by a two state stochastic approximate procedure.' *Proc. 6th IFAC Congress, Boston.*
23. R. L. Kashyap and A. R. Rao (1976). *Dynamic Stochastic Models from Empirical Data,* Academic Press, New York.
24. Load Forecasting Working Group (1980). 'Load forecast bibliography.' *IEEE Trans. on PAS,* **PAS-99**, iss. 1.

APPENDIX DERIVATION OF THE ZADEH–RAGAZZINI FILTER IN A PRACTICAL EXAMPLE

With regard to the load time series whose model is represented by equation (8.9), the design of the Zadeh–Ragazzini filter is based on the following steps:

(a) The functions $\varphi_s(\tau)$ and $\phi_s(\omega^2)$ are

$$\varphi_s(\tau) = K\,\mathrm{e}^{-|\tau|/\Theta} + H\,\mathrm{e}^{-\omega_n\zeta|\tau|}\cos(\omega_0|\tau| + \vartheta) = \sum_1^3{}_i X_i\,\mathrm{e}^{x_i} \tag{A8.1}$$

$$\phi_s(\omega^2) = -\sum_1^3{}_i \frac{2X_i x_i}{(\omega^2 + x_i^2)} = \frac{F(\omega^2)}{G(\omega^2)} \tag{A8.2}$$

with

$$\begin{aligned} &x_1 = 1/\Theta \qquad x_2 = -\omega_n\zeta + \mathrm{j}\omega_0 \qquad x_3 = x_2^* \\ &X_1 = K \qquad X_2 = \frac{H}{2}\mathrm{e}^{\mathrm{j}\vartheta} \qquad X_3 = X_2^* \end{aligned} \tag{A8.3}$$

where * means conjugate.

The polynomial $F(-p^2)$ becomes

$$F(-p^2) = (p-\beta_1)(p-\beta_2)(p-\beta_3)(p-\beta_4) \tag{A8.4}$$

and its zeros are such that

$$\beta_1 = -\beta_2 \qquad \beta_3 = -\beta_4 \qquad \beta_1 = \beta_3^* \qquad \beta_2 = \beta_4^*. \tag{A8.5}$$

(b) The optimal impulse response $h_\lambda(\tau)$ has the form

$$\begin{aligned} h_\lambda(\tau) = [u(\tau) - u(\tau - T)][&A_0 + A_1\tau + B_1\mathrm{e}^{\beta_1\tau} + B_2\mathrm{e}^{\beta_2\tau} \\ &+ B_3\mathrm{e}^{\beta_3\tau} + B_4\mathrm{e}^{\beta_4\tau} + C_1\,\delta(\tau) + D_1\,\delta(\tau - T)] \end{aligned} \tag{A8.6}$$

and it is directly obtained from (8.16) by setting $l = 1$, $n = 3$, $m = 2$, and

$$R(p)\int_{-\infty}^{+\infty} \frac{\phi_s(\omega^2)}{F(\omega^2)} R(-\mathrm{j}\omega)\,\mathrm{e}^{\mathrm{j}\omega(t+\lambda)}\,\mathrm{d}\omega = 0. \tag{A8.7}$$

(A8.7) turns out identically satisfied provided that $\varphi_s(\tau)$ is a finite sum of exponentials or, in equivalent terms, provided that the spectrum $\phi_s(\omega^2)$ is a rational form as in (A8.2). In fact, from the general expression

$$\frac{\phi_s(\omega^2)}{F(\omega^2)}R(-j\omega)=\frac{1}{R(j\omega)}=\sum_1^n{}_i\frac{Y_i}{(j\omega-x_i)} \qquad \text{(A8.8)}$$

where Y_i are the residuals corresponding to the poles x_i of $R(p)$, taking into account that

$$R(p)=\alpha_0+\alpha_1 p+\ldots+\alpha_n p^n=\sum_0^n{}_i\alpha_i p^i \qquad \text{(A8.9)}$$

the following relationships can be derived

$$\begin{aligned}R(p)\int_{-\infty}^{\infty}\frac{\phi_s(\omega^2)}{F(\omega^2)}R(-j\omega)\,e^{j\omega(\tau+\lambda)}\,d\omega&=\sum_0^n{}_i\alpha_i p^i\int_{-\infty}^{\infty}\sum_1^n{}_i\frac{Y_i}{(j\omega-x_i)}e^{j\omega(t+\lambda)}\,d\omega\\&=\sum_1^n{}_i Y_i(\alpha_0+\alpha_1 x_1+\ldots+\alpha_n x_i^n)\,e^{x_i(\tau+\lambda)}\\&=0 \qquad \text{(A8.10)}\end{aligned}$$

which proves (A8.7).

(c) Inserting $h_\lambda(\tau)$, as given by (A8.6), into (8.13) the following condition is obtained

$$\begin{aligned}&C_1\sum_1^3{}_iX_i\,e^{x_it}+D_1\sum_1^3{}_iX_i\,e^{x_i(\tau-t)}+\sum_1^3{}_i\{(X_i/x_i^3)\,e^{x_it}(A_0x_i^2+A_1x_i)\\&\quad+e^{x_i(T-t)}[A_0x_i^2+A_1(X_i^2T-x_i)]-2x_i^2(A_1t+A_0)\}+\sum_1^4{}_iB_i\,e^{\beta_it}\left[\sum_1^3{}_i\frac{2x_jX_j}{(X_j^2-x_j^2)}\right]\\&\quad+\sum_1^4{}_i\sum_1^3{}_j\frac{B_iX_j}{(\beta_i^2-x_j^2)}[-e^{x_it}(\beta_i+x_j)\,e^{\beta_iT}\,e^{x_j(T-t)}(\beta_i-x_j)]\\&\qquad=\mu_0+\mu_t+\sum_1^3{}_iX_i\,e^{x_i(t+\lambda)}. \qquad \text{(A8.11)}\end{aligned}$$

Since the third term of the left-hand side vanishes, (A8.11) gives rise to the following equations:

$$2A_0\sum_1^3{}_i(X_i/x_i)=\mu_0 \qquad \text{(A8.12)}$$

$$2A_1\sum_1^3{}_i(X_i/x_i)=\mu_1 \qquad \text{(A8.13)}$$

$$(1/x_i)A_0+(1/x_i^2)A_1+\sum_1^4{}_jB_j/(\beta_j-x_i)+C_1=e^{x_i\lambda} \qquad i=1,2,3 \qquad \text{(A8.14)}$$

$$(1/x_i)A_0+(1/x_i^2)(x_iT-1)A_1+\sum_1^4{}_j[B_j\,e^{\beta_jT}(\beta_j-x_i)]+D_1=0 \qquad i=1,2,3. \qquad \text{(A8.15)}$$

Moreover, inserting $h_\lambda(\tau)$ as given by (A8.6) into (8.14), two additional conditions are obtained:

$$A_0T+A_1T^2/2+\sum_{1}^{4}{}_i\,(B_i/\beta_i)(e^{\beta_iT}-1)+C_1+D_1=1 \qquad \text{(A8.16)}$$

$$A_0T^2/2+A_1T^3/3+\sum_{1\,i}^{4}(B_i/\beta_i^2)[e^{\beta_iT}(\beta_iT-1)+1]+D_1T=-\lambda. \qquad \text{(A8.17)}$$

(A8.14) and (A8.17) form a linear system of equations in the eight unknowns A_0, A_1, C_1, D_1, B_1, B_2, B_3, and B_4. The remaining (A8.12), (A8.13) yield the numerical values of the Lagrangian multipliers, after A_0 and A_1 have been replaced by the solutions obtained from such an equation system.

PART THREE: WIDER ISSUES

Comparative Models for Electrical Load Forecasting
Edited by D. W. Bunn and E. D. Farmer

CHAPTER 9

Two-step Modelling for Short-term Forecasting

Ramu Ramanathan, Clive Granger, and Robert Engle

1 INTRODUCTION

In modelling the demand for electricity, a utility or regulatory agency usually faces one of three time frames. In the very short run, the utility is concerned with choosing a generating mix, a maintenance schedule, and possibly emergency measures to meet the demand from hour to hour. In the intermediate run of one to five years, it is faced with the problem of meeting changing demand patterns with an essentially fixed capacity. In this situation, the role of demand forecasting models is to identify the need, if any, for power sharing arrangements, to set prices, or to take nonprice measures to match supply with demand. In the long run, such a model can help determine the optimal generating capacity and plant mix, pricing, and regulatory structure.

During the fifties and sixties, fuel prices, cost of capital, and regional economic and demographic trends were all stable. In those years, simple forecasting techniques such as trend extrapolation were adequate to project future peak loads, load factors, and load duration curves. Since the early seventies, the phenomenal rise in energy prices and the cost of capital has stimulated an increasing concern for conservation, alternative sources of energy, and for addressing a variety of load management issues. There is thus a great need for developing a modelling strategy that rests firmly on the underlying determinants of the demand for energy.

An approach to modelling energy demand that has been found to be very successful is the econometric method. Such an approach consists of formulating a statistical relationship between energy consumption and the determinants of demand (both short run determinants such as weather as well as long run determinants such as changing demographic and economic conditions), obtaining historical data, and then using multiple regression analysis to estimate the parameters of the model. If the data series covers a short period such as a month, then the estimated load curve has no reliable predictive power beyond that one month. If, as is often the case, the period was several years, then the average load curve could be meaningless because it is likely to have shifted over time. An appropriate modelling technique would be to obtain data for long periods of time and build into the model a means by which the shifts in the load curve may be estimated. In principle, this is simply the pooling of time series and cross section data from which Generalized Least Squares (GLS) estimates may be obtained. In practice, however, such a pooling leaves an investigator with a massive amount of data which require special attention to computational and econometric issues.

In response to the need for using large bodies of data in an efficient away, the authors developed a two-step procedure and applied it to residential survey data on a random sample of customers of the Connecticut Light and Power Co.[1] Using the same Connecticut data, Hendricks, Koenker, and Poirier[2] formulated a slightly different two-step approach. In a later study of hourly consumption for several regions in the Continental United States, Granger, Engle, and Ramanathan used the same approach with a good deal of success.[3,4] The present chapter discusses the two-step procedure and applies it to the two cases just mentioned. Section 2 describes the technique in detail and addresses the computational aspects also. Section 3 is the application of the methodology to residential hourly load curves. Section 4 presents the results of the analysis using regional system-wide hourly load and related data. Section 5 has some concluding remarks.

2 THE TWO-STEP METHODOLOGY

The electricity usage at a given hour of the day is determined by two types of variables. The first is related to the day of the week and the time of the day. This will include both life-style effects as well as environmental determinants. Those resulting from the life-style of a household are: the time that a husband and/or wife is at work and the children are at school, the types of activities enjoyed in the evenings and weekends, etc. The environmental causes include the levels of temperature, windspeed, and humidity, the number of hours of sunlight, the occurrence of public holidays, time-of-day pricing, if any, etc. All these variables change quickly from hour-to-hour or day-to-day. The second type of variable determining the hourly load is that which changes slowly over time: income, family size and composition, the size of a house, the type and number of appliance stocks, etc. If we are dealing with regional demand patterns rather than household demand, the list of variables would include population, income and employment levels, industrial mix of the area, its demographic attributes, appliance saturation levels, and the price of electricity.

The above dichotomy suggests a natural way of modelling the demand for electricity by time-of-day, whether at the residential level or at that of a system as a whole. First construct, for each individual or region, a short-term model (referred to as the 'time series' model) that relates hourly consumption to the quickly changing variables, viz. current and recent past weather conditions, time-of-day, and the day of the week. The time horizon for this short-term model would be a month or a quarter during which variables such as income, population demographics, appliance characteristics, and industrial mix may be treated as unchanged. The hourly load curve thus estimated will be the average over that month or quarter. Over time and across regions or individuals, the time series results will vary because of differences in appliance stocks and usage rates, and general economic, demographic, and industry conditions. The second stage of estimation (called the 'cross-section') thus relates the parameters of the first stage to these economic and demographic variables. The second stage, therefore, models how the short-run load curves vary over time and/or across regions. From these, forecasts of hourly loads can be obtained as follows. The economic and demographic variables, together with appliance saturations and electricity prices (all of which must be forecast independently), will be combined with the estimated second-stage parameters to generate estimates of the first-stage parameters. These will then be combined with the weather and calendar time to produce forecasts of the loads by the hour of the day. Using this framework, it is possible to carry out sensitivity analyses by changing weather scenarios or those on prices, appliance stocks, and utilization rates. Such an analysis can be useful in addressing some load management and conservation issues.

2.1 The time series model

For convenience of presentation, we assume that the time horizon for the short-run model is a quarter consisting of 2184 hourly observations (91 days and 24 hours for each day). Let $\mathbf{y}$ be the stacked vector of loads expressed in logarithms (which the authors have found to reduce heteroskedasticity), W be a $2184 \times K$ matrix of weather and other variables, and H be the 2184×24 matrix of hourly dummies. Each such dummy (i.e. binary) variable takes the value 1 at the appropriate hour and 0 elsewhere. The time series model may thus be written as follows:

$$\mathbf{y} = H\alpha + W\beta + \mathbf{u} \tag{9.1}$$

where $\mathbf{u}$ is the vector of residuals whose properties will be specified later.

The dimension of the observation matrix will be $2184 \times (K+24)$. The calculation of the cross-product matrix, its inversion, etc., will involve many computational steps. If we consider that several years' worth of data will typically be available, the whole operation will have to be repeated a large number of times. The computational burden of this procedure can therefore be enormous. To solve this, the authors exploited two properties of the H matrix. First, its columns are orthogonal to each other, and, secondly, the entire matrix is unchanged from quarter to quarter and region (household) to region (household). To see the usefulness of these properties note that the least-squares estimates of $\boldsymbol{\alpha}$ and $\boldsymbol{\beta}$ are given by

$$\begin{bmatrix}\hat{\alpha}\\ \hat{\beta}\end{bmatrix}=\begin{bmatrix}H'H & H'W\\ W'H & W'H\end{bmatrix}^{-1}\begin{bmatrix}H'\mathbf{y}\\ W'\mathbf{y}\end{bmatrix}.$$

Using the formula for the inverse of a partitioned matrix, we get

$$\begin{bmatrix}\hat{\alpha}\\ \hat{\beta}\end{bmatrix}=\begin{bmatrix}A & B\\ B' & C\end{bmatrix}\begin{bmatrix}H'\mathbf{y}\\ W'\mathbf{y}\end{bmatrix}$$

where

$$\begin{aligned}A^{-1}&=H'H-H'W(W'W)^{-1}W'H\\ B&=-(H'H)^{-1}H'WC\\ C^{-1}&=W'W-W'H(H'H)^{-1}H'W.\end{aligned}$$

Because the columns of the H matrix are orthogonal, $H'H=nI$, where I is the identity matrix of order 23 (one of the dummy variables will have to be omitted to avoid exact collinearity with the constant term), and n is the number of days for which the estimation is done. The above expressions simplify to the following:

$$\begin{aligned}A^{-1}&=nI-H'W(W'W)^{-1}W'H\\ B&=-H'WC/n\\ C^{-1}&=W'W-(W'HH'W/n).\end{aligned}$$

The matrices A, C, and $W'W$ have dimensions nearly half of the full cross-product matrix. It is well known that matrix inverses involve operations of the order of the cube of their dimensions and hence there will be a considerable reduction in computational costs if the matrices can be partitioned as above. For instance, the relative cost of inverting a 23×23 matrix as compared with a 55×55 matrix is only 7.3%. The overall reduction in costs can therefore be substantial.

With hourly data, the residual vector $\mathbf{u}$ in equation (9.1) is likely to be serially correlated. If it is AR(1) then we have $\mathbf{u}=\rho\mathbf{u}_{-1}+\boldsymbol{\varepsilon}$ where $\mathbf{u}_{-1}$ is the vector of lagged values of $\mathbf{u}$ and $\boldsymbol{\varepsilon}$ is white noise. The presence of serial correlation destroys the orthogonality exploited by the partitioned regression technique. To see this, rewrite the model as

$$\mathbf{y}^*=H^*\alpha+W^*\beta+\boldsymbol{\varepsilon} \tag{9.1a}$$

where $\mathbf{y}^*=\mathbf{y}-\rho\mathbf{y}_{-1}$ and similarly for H^* and W^*. Even though H is orthogonal, H^* is not and hence the gain in computational efficiency appears no longer feasible. Fortunately, the inherent linear restriction in the hourly dummies, viz. that all 24 hourly dummies add up to 1, enables us to preserve the orthogonality property (see Reference 5 for further explanation).

To summarize, the first stage of the two-step procedure consists of estimating the relationship between hourly loads and the short-run variables such as weather and hourly and day dummies. There will thus be one regression for each quarter, with associated regression coefficients for each right-hand side variable.

2.2 The cross-section model

The estimation problems in the cross-section stage are quite different from those in the time series model. There are as many cross-section regressions as there are explanatory variables in the time series model. The number of observations (N) will be determined by the number of quarters for which data are available multiplied by the number of cross-section units (households or regions). A typical cross-section model will be as follows:

$$\beta_j = \mathbf{Z}_j \boldsymbol{\gamma}_j + v_j \tag{9.2}$$

where β_j is the jth parameter of the first stage, $\boldsymbol{\gamma}_j$ is the vector of second-stage parameters, $\mathbf{Z}_j$ is the vector of explanatory variables in the cross-section, and v_j is the error term which may also be serially correlated. In the actual estimation of $\boldsymbol{\gamma}_j$ we do not observe β_j, but an estimate $(\hat{\beta}_j)$ which is subject to an error. Thus the estimated second-stage model is

$$\hat{\beta}_j = Z_j \boldsymbol{\gamma}_j + v_j + \eta_j = Z_j \boldsymbol{\gamma}_j + \delta_j \tag{9.3}$$

where $\eta_j = \hat{\beta}_j - \beta_j$ is the error in $\hat{\beta}_j$. The v_j do not have a constant variance and therefore the model has heteroskedasticity.

The appropriate technique in this case is a GLS procedure[6] which the authors used in a study on residential load curves and time-of-day pricing.[7] Assume that

$$E(v_j \eta_j') = 0 \qquad E(v_j v_j') = \sigma_j^2 I \qquad E(\eta_j \eta_j') = D_j.$$

D_j is a diagonal matrix of the estimated variances of the j regression coefficients for each of the cross-section observations. The covariance matrix of the disturbances of (9.3) is $\Omega_j = \sigma_j^2 I + D_j$. The unknown parameter σ_j^2 can be estimated from the least-squares residuals e_j of (9.1a). Hanushek[6] has shown that

$$E(e_j' e_j) = (N-L)\sigma_j^2 + \text{tr}\,[D_j - (Z_j' Z_j)^{-1}(Z_j' D_j Z_j)].$$

Therefore, an unbiased estimator of σ_j^2 is

$$\{e_j' e_j - \text{tr}\,[D_j - (Z_j' Z_j)^{-1}(Z_j' D_j Z_j^{-1})]\}/(N-L).$$

This procedure is easy to follow although the evaluation of the second term in the trace is not computationally trivial.[7] The matrix $\hat{\Omega}_j = \hat{\sigma}_j^2 I + D_j$ is a consistent estimator of the disturbance covariance matrix and thus GLS will be asymptotically efficient. The estimates are

$$\hat{\gamma}_j = (Z_j'\hat{\Omega}_j^{-1}Z_j)Z_j'\hat{\Omega}^{-1}\hat{\beta}_j. \tag{9.4}$$

In the next two sections, we present applications of the two-step procedure just described. One of them used data for a random sample of households in Connecticut and the other used data for 32 aggregate regions of the U.S.

3 RESIDUAL LOAD CURVES AND TIME-OF-DAY PRICING

The first application of the two-step procedure described in the previous section is based on a comprehensive data set collected by the Northeast Utilities under funding from the Federal Energy Administration and the Electric Power Research Institute (EPRI). In October 1975, a random sample of approximately 400 Connecticut residences were fitted with special meters that monitored, over the next year, the usage of electricity at quarter-hour intervals. Nearly 200 families belonged to the 'experimental group'. They faced a price structure that varied depending on the hour of the day the energy was consumed. During the peak hours (9 a.m. to 11 a.m. and 5 p.m. to 7 p.m. for winter weekdays and 10 a.m. to noon and 1 p.m. to 3 p.m. for summer weekdays), the rate per kilowatt-hour (kWh) of electricity was 16 cents. An intermediate rate of 3 cents per kWh prevailed during the hours 8 a.m. to 9 p.m. that were not in the peak period. A low rate of one cent per kWh was charged during all the other hours. On weekends and public holidays, the peak period was priced at the intermediate rate. The other group of households were used as 'control' and faced the same price at all hours of the day.

In addition to the usage data, information on the demographic and appliance mix characteristics of a family is available in substantial detail. Matching weather data is also available.[8] It is worthwhile mentioning that Burbank found that the households in the experimental group on average used over 25% less electricity during the peak hours than those in the control group.

3.1 The time series model

Because of missing data and other problems, many observations had to be eliminated from the analysis. Using the hourly observations, time series regressions were run separately for winter and summer months for 135 households of which 85 faced a peak pricing schedule. The remaining 50 households were in the 'control' group and faced a declining block price. To model the major regularities in electricity usage displayed by a household due to the habits of life-style requirements of its members, a set of hourly dummies were used. The hours 1 a.m. to 4 a.m. typically had very low usages and were excluded to act as a base-period against which other hours could be compared. In an earlier version of the study, day dummies had also been used but it was found that there were no significant differences in the shape of the load curve across weekdays. Weekend consumption patterns were naturally different. To allow for that, three variables were introduced; SIN2 and COS2 defined to be $\sin(\pi t/12)$ and $\cos(\pi t/12)$ during weekends and zero on other days, and the variable WORK which took the value 1 on working days and zero on other days,, including public holidays.

The time-of-day pricing effect was measured using two dummy variables:

$$\text{PEAK} = \begin{cases} 1 & \text{when peak prices were operating} \\ 0 & \text{at other times} \end{cases}$$

$$\text{HIGH} = \begin{cases} 1 & \text{when high-use prices were operating} \\ 0 & \text{at other times.} \end{cases}$$

To investigate the effects of school vacations, a variable SCHVAC was used. This variable took the value 1 during the vacation periods in winter and zero at other periods. Because the entire summer was a vacation period, no such variable was needed then.

The most important short-term causal variables are related to the weather. The process by which outside temperatures, windspeed, and humidity change the temperature inside a housing unit involves a very complicated transfer function, possibly including nonlinear terms. In an attempt to pick up the delayed effect of exceptional temperatures, the moving average of temperature over the preceding 24 hours, denoted by TEMPMA, was included as an explanatory variable. For the winter period, both the present temperature (TEMPNOW) and its squared value (TEMPSQ) were used. During the summer, the temperature itself may be less important than a measure of discomfort due to combinations of the temperature and humidity. A convenient proxy for the human discomfort index is the average of the wet and dry bulb temperatures (called TEMPHUM). After some experimentation with alternative formulations, WINDSPEED was added for both winter and summer periods. Based on preliminary analyses, solar radiation (calories per sq. cm per minute) and wind direction were omitted as being insignificant.

Table 9.1 presents the regression coefficients and t-value for a typical household for both summer (July and August 1976, excluding July 4) and winter (December 1975 to February 1976, excluding Christmas and New

TABLE 9.1 Time series regression results for particular households

Variable	Summer	Winter	Variable	Summer	Winter
HR1*	0.17 (0.7)	−0.29 (−1.8)	WORKDAY	0.19 (2.0)	−0.04 (−0.5)
HR5	0.27 (1.2)	−0.02 (−0.1)	PEAK	0.21 (0.8)	−0.26 (−1.5)
HR6	0.21 (0.9)	−0.21 (−1.3)	WINDSPEED	−0.016 (−0.75)	−0.006 (−0.6)
HR7	0.22 (1.0)	−0.11 (−0.6)	TEMPMA	0.18 (5.9)	−0.02 (−2.4)
HR8	0.18 (0.8)	0.41 (2.5)	TEMPHUM	0.08 (3.6)	
HR9	1.00 (4.3)	1.20 (6.6)	TEMPSQ	—	0.002 (2.4)
HR10	1.80 (7.6)	2.20 (9.9)	TEMPNOW	—	0.013 (1.6)
HR11	1.30 (4.7)	2.40 (11.1)	SIN2	0.06 (0.4)	0.27 (2.9)
HR12	0.93 (3.2)	1.60 (9.7)	COS2	−0.01 (−0.1)	−0.05 (−0.5)
HR13	0.79 (3.4)	1.10 (6.8)	CONSTANT	−4.68 (−8.8)	7.90 (6.8)
HR14	0.31 (1.1)	0.68 (4.3)	>		
HR15	0.54 (1.8)	0.55 (3.3)	Adj. R^2	0.203	0.331
HR16	0.80 (3.6)	0.66 (4.1)			
HR17	0.86 (3.7)	0.90 (5.3)			
HR18	1.20 (5.1)	2.00 (9.2)			
HR19	1.10 (4.9)	2.30 (10.4)			
HR20	1.10 (4.7)	2.00 (12.1)			
HR21	1.70 (7.3)	2.60 (15.5)			
HR22	1.20 (5.3)	1.20 (7.3)			
HR23	0.88 (3.9)	0.55 (3.6)			
HR24	0.52 (2.4)	0.56 (3.3)			
SCHVAC	—	0.35 (4.7)			

t-values are given in parentheses.
*HR1 is midnight to 1 a.m.

Year). The hourly dummies indicate active usage starting at 8 a.m., then a mid-morning peak, a lull in the afternoon, with the most intensive usage in the evening. The significance of SIN2 in the winter suggests a deviation between weekday and weekend patterns for that period. As can be expected, TEMPHUM is significant in summer and many temperature variables are significant in the winter.

Detailed results of the time series regressions are presented in Reference 9. Most of the results followed intuition. For instance, electricity consumption was negatively related to the current temperature during the winter, and in the summer the temperature–humidity index was positive and generally significant. The temperature moving average was also significant in many cases indicating that the temperature effects were spread over more than one hour. The sine and cosine variables were also significant, pointing out that behaviour was different during weekends. This difference was more pronounced in the winter.

Table 9.2 summarizes the significance of important explanatory variables. It presents the percentage of coefficients with *t*-statistics greater than 1.9 in

TABLE 9.2 Percentage of significant coefficients

Variable	Experimental group Summer		Experimental group Winter		Control group Summer		Control group Winter	
PEAK	57	(8)	56	(6)	39	(39)	49	(21)
TEMPNOW			47	(0)			55	(12)
TEMPSQ			29	(32)			24	(66)
TEMPHUM	33	(78)			50	(79)		
TEMPMA	49	(93)	57	(5)	50	(83)	55	(11)
WINDSPEED	28	(14)	40	(94)	29	(64)	39	(94)
SCHVAC			36	(81)			41	(80)
SIN2	36	(77)	63	(96)	23	(82)	67	(94)
COS2	38	(76)	48	(85)	35	(76)	55	(81)
WORK	32	(32)	48	(27)	38	(33)	36	(11)

Numbers in parentheses indicate per cent of these positive.

absolute value. To summarize the goodness of fit, households were classified into five equal subsamples according to the previous year's total usage; class 1 represented those in the lowest 20% of users and class 5 was the 20% heaviest users. The average R^2, adjusted for degrees of freedom, is presented in Table 9.3 for these groups. Although the very heaviest users, who typically used electricity for heating, had the highest R^2 values, in general there is no systematic relationship between the level of usage and goodness of fit. However, the R^2 values were consistently higher in the winter than in the summer, suggesting either more consistency of the daily shape (as picked up by the hourly dummies) in the winter or the greater explanatory power of the weather variables during the cold months.

3.2 Cross-sectional analysis

For each household, we were provided with data on 60 household-specific variables, as recorded in September 1976, which was just before the recording of electricity usage began. Many of these variables, however, could not be used because of collinearity problems or very rare occurrence in the sample. The variables used in the cross-section regression were the numbers of electric ranges, self-cleaning ovens, electric dryers, self-defrosting refrigerators, manual-defrosting refrigerators, freezers, dishwashers, colour TV sets, black-and-white televisions, humidifiers, dehumidifiers, window air conditioners, number of persons in the household in each of three groups (under 18, 18–64, and 65 or over), age of the house, square footage of the house, and the following dummy variables: one for the experimental pricing group, central air conditioning, electric main heating system, electric supplementary heating system, electric water heater, swim-

TABLE 9.3 Average R^2 values

	Class 1	2	3	4	5	Total
Experimental group						
Winter	0.315	0.281	0.360	0.352	0.507	0.363
Summer	0.259	0.235	0.281	0.260	0.272	0.261
Control group						
Winter	0.345	0.301	0.307	0.379	0.404	0.353
Summer	0.281	0.236	0.225	0.214	0.309	0.245

ming pool pump, type of structure, one for a single-family, and another for mobile homes. Based primarily on earlier experience, a subset of the above variables was chosen for each season.

Ordinary Least Squares (OLS) as well as GLS estimates were obtained for each of the time series regression coefficients in each season.[7,9] The GLS correction was different for each regression, but invariably it has little effect. The weights constructed varied by a maximum of a factor of two over the households, but frequently they varied only by a few per cent, because the homoskedastic component of the error term that is due directly to the 'noise' in the cross-sectional relation was much larger than the component due to uncertainty in the dependent variable. As the OLS and GLS results were so similar, we discuss only the OLS results. Tables 9.4 and 9.5 present, respectively, the summer and winter cross-sectional results. A variable is printed in italic type if $|t|>3$ and is given a query if the t-statistic is suggestive although not strictly significant (i.e. $|t|>1.6$). A negative sign in parentheses indicates that the coefficient in the cross-section regression is negative.

In general, the results seem promising. For some important variables the values of R^2 are adequate to good, and most of the significant explanatory variables make economic sense. A brief summary of the results follows for summer and winter. Details may be found in the reports already cited.

Summer results

The importance of certain appliances is clearly brought out in the hourly dummies. For instance, the electric water heater becomes a significant appliance in determining the regression coefficients for the hours 7 to 24 (Table 9.4). The dishwasher is significant for the hours 18 to 23. The electric dryer is significant from 11 p.m. to midnight, but less so in the preceding two hours. The interaction effect between air conditioning and the living area becomes significant at hours 12 and 14 to 23. Pool pump is important from

TABLE 9.4 Summer cross-sectional results for selected variables (OLS)

Dependent variable	Adj. R^2	d	Main significant household characteristics
CONSTANT	0.15	1.72	*Central air conditioning × square footage* (−), window air conditioning × square footage (−)
HR5[a]	0.03	2.02	Dehumidifier (−), window air conditioning × square footage, people under 18 (−?)
HR6	−0.03	2.02	Electric water heater (?)
HR7	0.02	1.72	Electric water heater
HR8	0.05	1.64	*Electric water heater*
HR9	0.06	1.69	*Electric water heater*, people 18–64
HR10	0.10	2.01	*Electric water heater*, people under 18, dehumidifier (−?), dishwasher (+?)
HR11	0.18	1.82	*Electric water heater, people under 18*, central air conditioning × square footage (+?)
HR12	0.18	1.61	*Electric water heater*, people under 18, central air conditioning × square footage
HR13	0.13	2.11	*Electric water heater, people under 18*
HR14	0.19	1.65	*Electric water heater, people under 18, central air conditioning × square footage*
HR15	0.22	1.61	Electric water heater, *people under 18, central air conditioning × square footage*
HR16	0.18	1.95	Electric water heater, *people under 18, central air conditioning × square footage*
HR17	0.22	2.00	Electric water heater, *people under 18, central air conditioning × square footage*, heated pool, people 18–64 (+?)
HR18	0.24	1.95	Electric water heater, people 18–64, *people under 18*, heated pool, *central air conditioning × square footage*, people 65 or over (+?), dishwasher (+?)
HR19	0.28	2.02	*Electric water heater, people under 18*, people 18–64, *central air conditioning × square footage, dishwasher*
HR20	0.28	1.86	*Electric water heater, people under 18*, people 18–64, *central air conditioning × square footage*, dishwasher
HR21	0.26	2.04	Electric water heater, *people under 18*, people 18–24, *central air conditioning × square footage*, dishwasher
HR22	0.25	2.05	Electric water heater, people under 18, people 18–64, central air conditioning × square footage, dishwasher, electric dryer (+ ?)
HR23	0.26	1.93	Electric water heater, people under 18, people 18–64, central air conditioning × square footage, dishwasher, electric dryer, survey dummy (+?)[b]
HR24	0.17	1.94	Electric water heater, people under 18, people 18–64, electric dryer

TABLE 9.4—(*continued*)

HR4	0.12	1.96	People under 18, people 18–64
PEAK PRICE	0.20	2.01	People under 18 (−), *swimming pool pump* (−), *survey dummy* (−), central air conditioning×square footage (−?)
WEEKDAY	−0.07	1.84	None
WIND SPEED	0.06	2.07	Dehumidifier (−), survey (−)
TEMPERATURE MOVING AVERAGE	0.08	1.73	*Central air conditioning×square footage*, window air conditioning×square footage
TEMPERATURE HUMIDITY INDEX	0.16	2.09	*Central air conditioning×square footage*, window air conditioning×square footage
SIN2	−0.01	1.80	None
COS2	−0.01	2.05	People 18–64
HIGH HOURS	0.26	1.76	*People under 18, central air conditioning×square footage, electric water heater*, people 18–64
SUMMER TEMP	0.19	1.79	*Central air conditioning×square footage, window air conditioning×square footage*

[a] When interpreting hourly dummies it should be remembered that they are relative to electricity usage in the suppressed hours 2 to 4 (1 to 4 a.m.).
[b] Survey dummy = 1 if in experimental group, 0 otherwise.

TABLE 9.5 Winter cross-sectional results for selected variables (OLS)

Dependent variable	Adj. R^2	d	Main significant household characteristics
CONSTANT	0.68	1.88	Electric water heater, *electric heat×square footage*
HR5	0.12	2.07	*Electric heat×square footage*, people 18–64 (−)
HR6	0.11	2.15	Electric heat×square footage, supplementary electric heat, electric water heater (+?)
HR7	0.14	2.16	*Electric heat×square footage*, electric water heater
HR8	0.17	2.33	*Electric heat×square footage*, survey dummy (?)
HR9	0.18	2.34	*Electric heat×square footage*, dishwasher (?)
HR10	0.18	2.14	*Electric heat×square footage*, dishwasher
HR11	0.09	2.25	Electric heat×square footage (?), dishwasher (?), electric water heater (?), people under 18 (?)
HR12	−0.02	2.36	None
HR13	0.01	2.05	People under 18
HR14	0.02	2.10	People under 18
HR15	0.05	2.31	Electric heat×square footage (−), people under 18
HR16	0.03	2.29	People under 18 (?)

TABLE 9.5—(*continued*)

HR17	0.07	2.41	People under 18
HR18	0.19	2.28	Electric heat × square footage, people under 18, dishwasher
HR19	0.22	2.17	*People under 18,* dishwasher, electric water heater (?)
HR20	0.17	2.28	*People under 18,* dishwasher, electric water heater (?)
HR21	0.31	2.06	*People under 18,* dishwasher, electric heat × square footage, electric water heater (?)
MR22	0.31	2.04	*People under 18, dishwasher,* electric water heater, survey, electric heat × square footage (?)
HR23	0.28	2.19	People under 18, dishwasher, *electric water heater,* survey
HR24	0.18	2.08	People under 18, dishwasher, electric water heater, survey
HR1	0.10	2.12	People under 18, electric water heater, survey
PEAK PRICE	0.31	1.92	*Electric heat × square footage* (−), dishwasher (−), *survey* (−)
SCHVAC	−0.02	1.66	None
WEEKDAY	−0.05	2.26	None
WIND SPEED	0.54	2.39	*Electric heat × square footage,* electric water heater (?)
TEMPERATURE MOVING AVERAGE	0.77	1.83	*Electric heat × square footage* (−), electric water heater (−?)
TEMPERATURE	−0.00	2.20	Survey (−)
SQUARED	0.74	2.24	*Electric heat × square footage*
TEMPERATURE NOW SIN2	0.32	2.30	*Electric heat × square footage, people under 18,* people 18–64, dishwasher, electric water heater (?)
COS2	0.15	2.48	*Electric heat × square footage,* people under 18 (?)
HIGH HOURS	0.12	2.24	People under 18, electric water heater (?), dishwasher
SUM OF TEMPERATURES	0.83	2.09	*Electric heat × square footage* (−), supplementary electric heater (?)

4 to 5 p.m. The number of people under 18 is important at all times except during the hours 5 to 8.

To economists, the most interesting variable is the peak price. Not surprisingly, the experimental households showed a strongly significant negative regression coefficient, indicating that they used less electricity during the peak pricing period as compared with the control group. Households with a pool pump tended to shift out of the peak to reduce the electricity consumption during that period. Central air conditioning interacted with the area of the house so as to yield the expected negative effect on the peak price. Air conditioning times the area was the only variable that had any significant impact on the temperature variables. For windspeed, the dehumidifier had a strong negative effect.

Winter results

As in the summer, the hourly dummies exhibited interesting results. The dishwasher had a significant effect from 9 to 11 a.m. and again from 6 p.m. to midnight. The electric water heater had a noticeable effect from 6 to 7 a.m. and 9 p.m. to 1 a.m. In contrast to the summer results, the experimental group dummy variable was significant during the hours 9 p.m. to 1 a.m. With regard to the peak price, dishwasher, electric heating times area, and the survey dummy all had the expected negative effects which were statistically significant. Unlike the summer, the weather variables generally had much better fits in the winter. Electric heating times the area was significant in all cases. Electric water heater came close to being significant.

Because the prices did not vary across households (an unfortunate feature in the design of the study), it would not be sensible to attempt to derive a price elasticity. However, it is possible to infer which appliances are responsible for any shift away from the peak or for a curtailment of demand.[7] A substantial decrease in usage at the peak hours is clearly evident. There is also a spreading of the load, particularly into the evening and in the morning hours before peak prices come into effect. These shifts are associated particularly with households having electric heating but also with dishwashers and swimming pool pumps.

All the above results seem both sensible and persuasive and appear to indicate that the two-step procedure is indeed viable. Further analyses along these lines are clearly needed both to improve the ability to distinguish the critical determinants of residential load curves as well as to examine the response to time-of-day pricing.

4 AN APPLICATION TO REGIONAL SYSTEM LOAD CURVES

It is the experience of utilities across the country that over the years the system hourly load curve for a particular day of the year has changed both in level and in shape. Figure 9.1 presents an illustration of this for New Mexico. It is interesting to note that the 1966 load curve lay partly above that in 1970 even though there has been economic growth in the interim. It will also be noted that the shape itself is quite different between the early and later years. In a study sponsored by EPRI, the authors applied the two-step procedure to explain differences in the hourly load curves across 32 regions in the United States and over a period of 13 years.[3,4] The data were

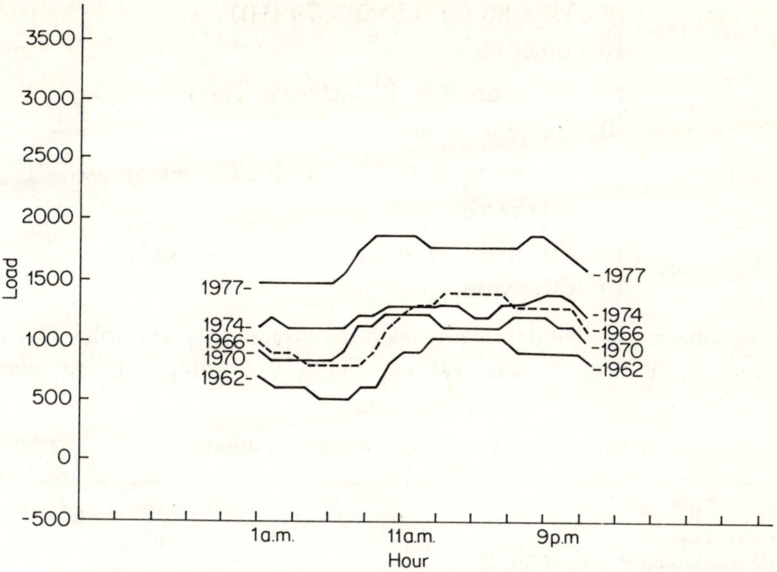

FIGURE 9.1 Observed hourly load curves on selected Tuesdays in summer (New Mexico)

provided by Data Resources Inc. (DRI) and consisted of the hourly load, matching weather information, and a number of region-specific cross-section variables for the period 1962–74 for the 32 regions listed in Table 9.6.

4.1 Time series model results

As in the study of households, there were hourly dummies with the period 2 a.m. to 3 a.m. excluded. There were also day dummies, one for each day except Wednesday. A day was defined to start at 3 a.m. For weekends, there were sine and cosine terms separately for Saturdays and Sundays. Two harmonics were used with periods 12 and 24 hours. Based on experimentation, a linear time trend was found to be useful.

The temperature variable was found to enter the model not only nonlinearly, but it interacted with the hour of the day. Thus the following cross-impact terms between the time-of-day and temperature ranges were used:

TEMP = temperature in degrees Fahrenheit

$$\text{LO} = \begin{cases} 60 - \text{TEMP} & \text{for TEMP} < 61 \\ 0 & \text{otherwise} \end{cases}$$

$$\text{MI} = \begin{cases} \text{TEMP} - 60 & \text{for } 60 < \text{TEMP} < 80 \text{ (Middle)} \\ 0 & \text{otherwise} \end{cases}$$

$$HI = \begin{cases} TEMP - 80 & \text{for } TEMP > 79 \text{ (HI)} \\ 0 & \text{otherwise} \end{cases}$$

$$D = \begin{cases} 1 & \text{for hours 9 to 17 inclusive (Day)} \\ 0 & \text{for other hours} \end{cases}$$

$$E = \begin{cases} 1 & \text{for hours 18 to 22 inclusive (Evening)} \\ 0 & \text{otherwise} \end{cases}$$

$$N = \begin{cases} 1 & \text{for hours 23 and 24 and 1 to 8 (Night)} \\ 0 & \text{otherwise.} \end{cases}$$

All the nine cross-products were used as explanatory variables. To capture some of the lagged temperature effects, a 24-hour moving average

TABLE 9.6 List of regions used in the study

1. New England
2. New York
3. Pennsylvania–Jersey–Maryland
4. West Virginia
5. Virginia
6. Ohio 1—American Electric Power System
7. Ohio 2—Remaining
8. Indiana
9. Michigan
10. Wisconsin
11. Illinois
12. Kentucky
13. Tennessee
14. North and South Carolina
15. Alabama, Georgia
16. Florida
17. Minnesota
18. Iowa
19. Missouri
20. Louisiana
21. Oklahoma
22. North and South Dakota
23. Montana, Wyoming
24. Washington, Oregon, Idaho
25. Nebraska
26. Kansas
27. New Mexico
28. Colorado
29. Utah
30. Arizona
31. Nevada
32. California

(TEMPMA) of lagged temperature (excluding the current temperature) and a moving average (TEMPMN) of the past five midnight temperatures (not including the current temperature) were used. The effects of humidity were captured with three variables:

$$\text{RELHUM} = \text{relative humidity}$$

$$\text{HUMTLD} = \begin{cases} \text{RELHUM} \times (\text{60-TEMP}) & \text{for TEMP} < 61 \\ 0 & \text{for other temperatures} \end{cases}$$

$$\text{COMFORT} = \text{MAX}(\text{THI-70}, 0) \text{ where}$$

$$\text{THI} = \text{TEMP} - 0.55\ (\text{1-RELHUM/100})\ (\text{TEMP-58}).$$

The comfort index is a measure of the discomfort endured by individuals in high humidity and temperature situations.[10] The only other weather variable used in the time series model was WINDSPEED.

The detailed results of the time series model may be found in Reference 3. In general, the model fitted well with a majority of adjusted R^2 values over 0.9. These values were typically lower in the autumn and spring quarters, presumably because of extreme variations in temperatures during these periods. Table 9.7 shows the frequency distribution of R^2 values for six regions which represent a variety of climatic and socio-economic conditions.

Figures 9.2–9.4 represent the plots for selected regression coefficients from the California time series model. It is evident from these plots that the hourly load shape has not been stable over time.

The constant term in the time series regression represents the base consumption between 2 a.m. and 3 a.m. on Wednesdays. Changes in this are mostly due to the size and mix of the regional economy, and are generally not weather sensitive, nor are they affected by the day of the week. For the winter and spring quarters, there has been a steady rise in the constant term with only minor fluctuations. In contrast, the summer and autumn seasons had wide fluctuations but an underlying upward trend is still perceptible. It is interesting to note that between 1971 and 1972, and again between 1973 and 1974, there were reductions in the values in all the seasons. This may be partly due to a conservation effort following the Arab oil embargo and subsequent steep increases in the price of energy.

Almost all the hourly binaries were statistically significant indicating that the consumption patterns at all the hours were significantly different from those at the control hour. As can be seen from Figure 9.3, the hourly binaries showed clear downward trends in all the seasons. This indicates a flattening of the base load curve, at least in percentage terms.

The temperature variables and the various interactions with day, evening, or night-time binaries seem to have adequately captured the underlying load shape, except for days with unusually low temperatures. Each variable

TABLE 9.7 Frequency distribution of adjusted R^2

Quarter	<0.75	0.75–0.80	0.80–0.85	0.85–0.90	0.90–0.95	>0.95
(a) New England (Region 1)						
Winter	0	0	0	0	13	0
Spring	0	0	0	0	13	0
Summer	0	0	0	0	10	3
Autumn	0	0	0	0	13	0
(b) Alabama/Georgia (Region 15)						
Winter	2	0	0	1	10	0
Spring	0	0	0	4	9	0
Spring	0	0	0	0	1	12
Autumn	0	0	0	10	3	0
(c) Minnesota (Region 17)						
Winter	0	0	0	0	7	6
Spring	0	0	0	1	12	0
Summer	0	0	0	0	10	3
Autumn	0	0	0	4	9	0
(d) Washington/Oregon/Idaho (Region 24)						
Winter	8	0	0	0	3	2
Spring	0	0	0	2	11	0
Summer	0	0	0	0	12	0
Autumn	0	0	0	1	12	0
(e) Kansas (Region 26)						
Winter	0	0	0	0	13	0
Spring	0	0	0	5	8	0
Summer	0	0	0	0	9	4
Autumn	0	0	0	5	8	0
(f) California (Region 32)						
Winter	0	0	0	0	7	6
Spring	0	0	0	0	13	0
Summer	0	0	0	0	0	13
Autumn	0	0	1	0	12	0

included was significant in an overwhelming majority of regions. The time series regression coefficients for Comfort Index for summer and TEMPLON (low night-time temperature) in the winter are plotted in Figure 9.4 for California. These coefficients varied a great deal, but from 1967 to 1973 there has been a dramatic increase in the values for the Comfort Index, possibly due to an increase in the air-conditioning demand.

With a view to explaining the trends and fluctuations in the regression coefficients, the second stage of the two-step procedure related those values to the socio-economic characteristics of the regions. The results of that analysis are discussed in the next section.

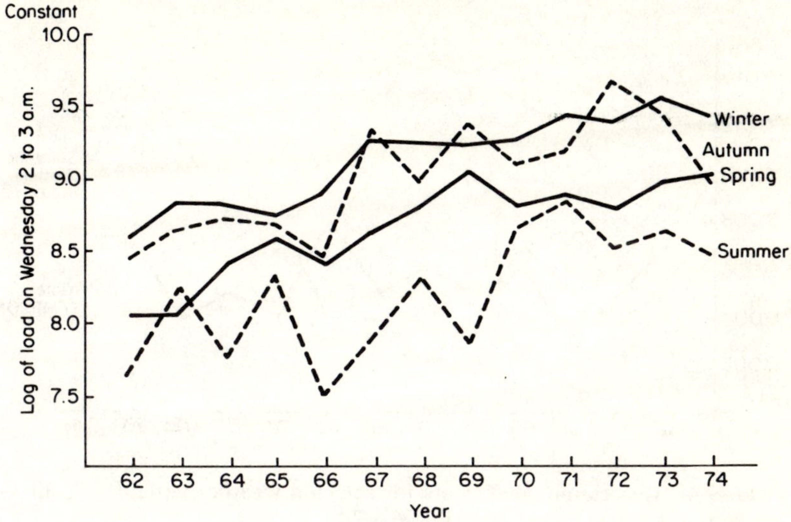

FIGURE 9.2 Estimated constant term for California (1962–74)

4.2 The cross-section model

As in the residential load curve model, each first stage regression coefficient is expressed as a function of the socio-economic variables. As it would be inappropriate to consider each region as a random observation from the same distribution, we specified different intercepts. This was accomplished by introducing regional binary variables, which took the value 1 for a

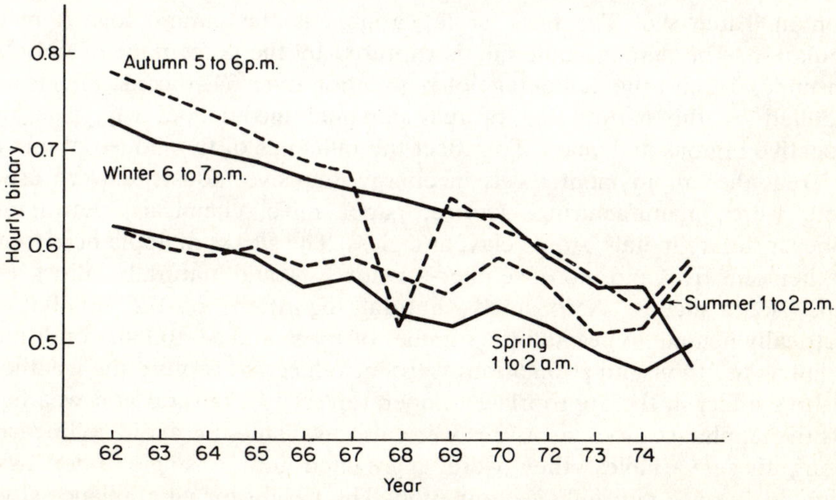

FIGURE 9.3 Selected hourly binaries for California (1962–74)

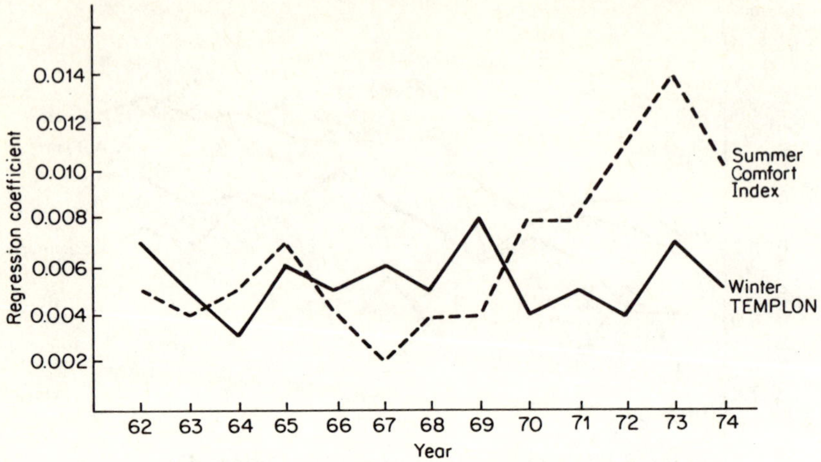

FIGURE 9.4 Regression coefficients for selected weather variables, California (1962–74)

specific region and zero elsewhere. The cross-section variables were quarterly for the 13 years 1962–74. The behaviour appears to be quite different from one season to another. This was true not only for the intercept but also for the slope coefficients. This is perfectly natural as the response of the load to weather depends very much on the season, and many other components of demand may depend on factors like the school year. The cross-section model was thus estimated separately for each season.

A complete list of the explanatory variables used in the second stage is given in Table 9.8. The basic scale variable is the natural logarithm of population. The demographic mix is captured by the percentage of population under 18 and the percentage of population over 64. Income effects are modelled by the natural log of real personal income per capita in the respective regions and years. To reflect the influence of the industrial mix in the area, the employment levels in energy-intensive industries were used. These were: manufacturing, service, paper mills, chemicals, petroleum, rubber, primary metals, stone, clay, and glass. The stocks of appliances (both weather-sensitive and non weather-sensitive) would naturally affect the hourly load shapes. Accordingly, natural logarithms of the number of electrically heated homes and the number of room and central air conditioners (all as per thousand population) were introduced to capture the weather-sensitive effect in the appropriate temperature range. Several non-weather-sensitive appliance stock measures were also available. To avoid collinearity among these variables, they were aggregated into a single index as a weighted average per million population. The weight for an appliance stock was based on the annual average usage in kWh for that appliance. The

TABLE 9.8 List of cross-section variables

No.	Mnemonic	Description
0	Const	Constant term
1–32	Region XX	Binary variable for region no. XX
33	POP	Natural log of population in millions
34	APLAGRE	Natural log of per thousand population appliance index
35	E MANUF	Employment in manufacturing per thousand population
36	E SERVICE	Employment in service per thousand population
37	E 26	Employment in paper mills per thousand population
38	E 28	Employment in chemicals per thousand population
39	E 29	Employment in petroleum per thousand population
40	E 30	Employment in rubber per thousand population
41	E 32	Employment in stone, clay, and glass per thousand population
42	E 33	Employment in primary metals per thousand population
43	APR	Natural log of average residential price in real $/kWh
44	API	Natural log of average industrial price in real $/kWh
45	YP	Natural log of real personal income per capita in $1000's
46	POP<18	Percentage of POP<18
47	POP>64	Percentage of POP>64
48	ELECHEAT	Natural log of the stock of electrically heated homes per 1000 population
49	ROOMAIR	Natural log of the stock of room air conditioners per 1000 population
50	CENTAIR	Natural log of the stock of central air conditioners per 1000 population

appliances used were refirgerators, freezers, electric ranges, electric water heaters, automatic and conventional washers, electric dryers, and dishwashers.

Table 9.9 shows the adjusted R^2 values for each of the hourly dummies. Similar values for the remaining time series variables are shown in Table 9.10. With some exceptions, the cross-section equations had good fits. It will be noted that the binaries for Tuesday and Thursday were mostly insignificant, indicating that the load shapes in these days were similar to those on Wednesdays. Midnight temperatures were also insignificant, presumably due to multicollinearity problems with the temperature moving average. Other

TABLE 9.9 Adjusted R^2 for the hourly binarycross-section regressions

First-stage		Season			
variable no.	Hour	Winter	Spring	Summer	Autumn
2	1	0.504	0.611	0.688	0.682
3	2	0.477	0.681	0.680	0.700
4	4	0.792	0.777	0.642	0.835
5	5	0.895	0.856	0.828	0.889
6	6	0.936	0.859	0.804	0.906
7	7	0.930	0.886	0.834	0.898
8	8	0.954	0.926	0.899	0.937
9	9	0.943	0.953	0.944	0.944
10	10	0.945	0.949	0.938	0.948
11	11	0.944	0.953	0.935	0.946
12	12	0.945	0.956	0.934	0.951
13	13	0.933	0.924	0.885	0.937
14	14	0.942	0.952	0.930	0.942
15	15	0.941	0.946	0.922	0.933
16	16	0.940	0.942	0.917	0.939
17	17	0.948	0.945	0.924	0.955
18	18	0.918	0.926	0.908	0.934
19	19	0.932	0.913	0.880	0.932
20	20	0.926	0.928	0.916	0.936
21	21	0.910	0.922	0.928	0.924
22	22	0.901	0.900	0.886	0.915
23	23	0.898	0.805	0.750	0.906
24	24	0.810	0.721	0.720	0.857

insignificant variables were uncommon in occurrence (for example, humidity—low in the summer). Every cross-section variable included was statistically significant in several regressions proving that their inclusion was worthwhile.

For a more detailed analysis of the cross-section results, the reader is referred to Reference 3. We discuss these results only briefly here.

The logarithm of population was used as a scale variable for explaining the variations in the constant term, which represents base consumption. In all the seasons, population was very significantly positive. Across the quarters the population elasticities varied from 0.8 to 1.8. It should be pointed out, however, that because of multicollinearity, this term may be capturing some effects from the stock of weather-sensitive and nonweather-sensitive appliances, as well as some from the price of electricity.

The percentage of population under 18 years of age affected several hourly binaries in a statistically significant way. Many of these coefficients were negative over all the seasons, indicating that regions with a higher fraction of the population under 18 tended to have a lower level of base

TABLE 9.10 Adjusted R^2 for remaining cross-section regressions

First-stage		Season			
variable no.	Name	Winter	Spring	Summer	Autumn
1	CONSTANT	0.995	0.990	0.964	0.989
25	MONDAY BINARY	0.471	0.366	0.234	0.372
26	TUESDAY BINDARY	0.069	0.077	0.002	0.006
27	THURSDAY BINARY	0.042	0.062	−0.024	0.090
28	FRIDAY BINARY	0.371	0.437	0.307	0.432
29	SATURDAY BINARY	0.921	0.929	0.925	0.912
30	SUNDARY BINARY	0.960	0.956	0.957	0.960
31	SATSIN1	0.890	0.866	0.845	0.883
32	SATCOS1	0.883	0.879	0.858	0.865
33	SATSIN2	0.713	0.694	0.734	0.741
34	SATCOS2	0.901	0.812	0.766	0.872
35	SUNSIN1	0.953	0.928	0.918	0.933
36	SUNCOS1	0.874	0.885	0.873	0.855
37	SUNSIN2	0.908	0.867	0.870	0.889
38	SUNCOS2	0.921	0.874	0.811	0.911
39	MONHR3–6	0.751	0.787	0.742	0.698
40	TIME	0.491	0.573	0.594	0.282
41	WINDSPEED	0.439	0.328	0.370	0.303
42	REL HUMIDITY	0.534	0.607	0.596	0.473
43	T LOW DAY	0.730	0.587	0.233	0.606
44	T LOW EVE	0.688	0.618	0.227	0.558
45	T LOW NIGHT	0.714	0.588	−0.015	0.612
46	T MED DAY	0.124	0.610	0.416	0.522
47	T MED EVE	0.119	0.313	0.305	0.631
48	T MED NIGHT	0.183	0.552	0.682	0.047
49	T HIGH DAY	N.E.*	0.160	0.398	−0.067
50	T HIGH EVE	0.384	0.313	0.505	0.102
51	T HIGH NIGHT	N.E.*	0.227	0.321	N.E.
52	HUMIDITY LOW TEMP	0.484	0.524	−0.011	0.463
53	COMFORT INDEX	0.540	0.503	0.768	0.000
54	TEMP MOVING AVE	0.636	0.830	0.835	0.531
55	TEMP MIDNIGHT	0.139	0.200	0.363	0.105

* N.E. means 'not estimated' because of the non-occurrence of the specified variable.

consumption, especially in the mornings and early in the evening. This may be because young children get up late and go to bed early and may also make less use of stereo, TV, etc. In contrast to this group, the percentage of population over 64 had significant positive effects on the hourly binaries. This effect is reinforced by the winter weather variables which had strong positive coefficients. Thus regions with more older people are also, not surprisingly, weather sensitive, with higher loads when the temperature is low.

As income rises, other things being equal, we would expect electricity consumption to rise also. The effect of per capita income on the constant term was significantly negative for summer and insignificant in the remaining seasons. One possible reason might be the multicollinearity between income and appliance stocks, which tends to obscure the separate effects. A more positive explanation is that high-income people might go on more extended vacations in the summer than low-income people, thus leaving their house vacant during that period. This would tend to depress the base load, on average. It should be noted, however, that because income has significant positive effects in many of the hourly binaries, and some weather-sensitive variables as well, the combined effect of income might be positive. In fact, subsequent simulations with the complete model indicated that such was indeed the case. It is therefore important to estimate income elasticities not from these partial regression coefficients but from total simulation effects.

The results for industrial employment were mixed. For some industries the cross-section coefficients were significantly negative in all seasons, but for others the coefficients were strongly positive. It is thus clear that while industrial mix, as measured by the percentage of employment, is important in affecting base load, there is no uniformity among industries in the nature of those effects.

The cross-section model included, as explanatory variables, the real average price of electricity for residential and industrial customers. Because these two series are highly correlated, it would be more appropriate to combine their effects. Such combined effects were consistently negative, although the individual partial effects sometimes had counter-intuitive signs.

The aggregate appliance index had the expected positive coefficient (with only minor exceptions) during all the hours of day and all the seasons. The weather-sensitive appliances (electric heaters, central and room air conditioners), however, had negative signs for most of the hourly binaries. Because the constant term and some of the temperature interactions were positive, the overall effect was actually positive from the simulations. This, once again, shows that an examination of only the partial effects could lead to very misleading interpretations.

4.3 Post-sample validation of the model

In order to check the quality of the forecasts and discover simple ways of improving the forecasts, the model estimated using the data for the period 1962–1974 was used to generate post sample period forecasts for the period 1975 to 1977 for the six test regions, New York, Wisconsin, New Mexico, Washington–Oregon–Idaho, Alabama–Georgia, and California. In many respects, the forecast period was far from perfect. The main impact of the first energy crisis started to occur in 1974–75, and hence a model constructed with data mostly prior to that period might be expected to perform poorly. The model did, however, perform well overall. Over the whole validation period, the mean errors for the first four regions were no more than 3.5%. For California and Alabama–Georgia, however, they were −17 and 15% respectively. Subsequent to this validation exercise, it was discovered that one of the California utilities stopped reporting in the validation period. If a crude adjustment of 13% (the estimated size of this utility) is made, California's average error drops to about 4.5%, which is reasonable. For the Alabama–Georgia region, however, there seemed to be no such data problem and no adequate explanation is available for the model's poor performance.

The variances of the forecasts were uniformly small, indicating that our model captured the shape of the load curve very well. In summary, five of the six test regions gave good results, but the sixth did not. Considering that 1974–77 was an unusual period, the overall validation results are reassuring for the two-step methodology.

4.4 Sensitivity analysis

The model described in the previous sections was re-estimated after including the data from 1975 to 1977. Slight adjustments to the specification were made, to allow for problems such as the nonreporting of a utility in California. The re-estimated model was then used to generate illustrative forecasts under a variety of scenarios.[4]

In order to assess the sensitivity of these results to the assumptions, elasticities of the load with respect to the input variables were obtained for the six test regions. All these were constructed as perturbations around estimates for the year 2000 with 1975 weather. These elasticities did not vary across regions. The weather-sensitive appliances (central air conditioners, electric heating, and room air conditioning) all had elasticities of roughly 6 per cent. In other words, a doubling in the saturation of one of these would give a 6 per cent increase in the average system load. The aggregate of all non weather-sensitive appliances had a much larger effect at 0.4 while per capita income and price elasticities were 0.2 and −0.2, respectively.

Finally, the sensitivity to the weather variables was examined, using 1976 and 1977 weather. The results showed substantial differences, particularly at the peak. For one of the regions, the annual peak differed by 25 per cent from the highest to the lowest of the weather patterns. Recalculation of the above elasticities with the 1977 weather clearly showed that even the elasticities were somewhat sensitive to the weather.

5 CONCLUDING REMARKS

This chapter has presented two applications of the two-step methodology for forecasting hourly load curves. The very complex models have been tested in many different ways and, in most cases, have come out with sensible and defensible conclusions. The forecasts and sensitivities to changes in assumptions, generated in the Regional Load Curve Study, are the kinds of basic raw materials upon which utilities must make long-range planning decisions.

In what way does our approach differ from other studies on time-of-day demand for electricity? First, the model has been tested on post-sample data and found to be reasonably accurate. Second, the Regional Load Curve model has used all the information available from the experience of a wide variety of regions on the growth of loads and load shapes to forecast future patterns. Insofar as some of the regions experienced patterns which others will face in the future, this structure is already included in the model. Third, the model analyses the entire load rather than some portions in detail leaving a large segment under a miscellaneous category whose behaviour it would be extremely hard to forecast. Finally, this model captures the weather-sensitive load very well so that changes in the base load and load shape can be examined without the noise due to weather fluctuations.

Although the model has a number of positive attributes, its limitations should also be borne in mind. In any analysis of the type undertaken here, uncertainty enters in a number of ways. First, there is the inherent uncertainty in any statistical procedure. Confidence intervals for the regression coefficients give some indication of the accuracy with which they are measured. In our models, there are so many regression coefficients, both first and second stage, that it would be impossible to measure the overall accuracy of a prediction based upon the joint distribution of the coefficients. It is, however, always true that the further the forecast point is from the historical data, the more uncertain the results.

Another source of uncertainty is the misspecification of the model. Suppose the second-stage regression for a particular variable does not have a good fit, possibly because the model is misspecified. In order to generate forecasts by the hour, we must necessarily forecast the first-stage parameter whose fit was not good, and hence whose value is not accurate. Improved specification of the model could reduce this uncertainty, but it is impossible to eliminate it altogether.

The modelling approach proposed here is only a beginning and should be considered a preliminary source of forecasts. With further careful analysis, the accuracy of hourly load forecasts, both in the short and long runs, can yet be substantially improved.

REFERENCES

1. C. W. J. Granger, A. Andersen, R. F. Engle, and R. Ramanathan (1977). 'Long term residential load forecasting.' Anthony Lawrence (Ed.), *Forecasting and modelling time-of-day and seasonal electricity demands*, EPRI, EA 578-SR.

2. W. Hendricks, R. Koenker, and D. J. Poirier (1977). 'Residential demand for electricity: An econometric approach.' Anthony Lawrence (Ed.), *Forecasting and modelling time-of-day and seasonal electricity demands*, EPRI, EA 578–SR.
3. R. Ramanathan, R. F. Engle, and C. W. J. Granger (1981). 'Regional load-curve models: QUERI'S model specification, estimation and validation.' EPRI, EA 1672, Vol. 2.
4. R. F. Engle, C. W. J. Granger, and R. Ramanathan (1981). 'Regional load-curve models: QUERI'S model long-run forecasts and sensitivity analysis.' EPRI, EA 1672, Vol. 4.
5. R. Ramanthan and A. Mitchem (1982). 'Econometric and computational issues in estimating demand for energy by time-of-day.' *Rev. of Econ. and Stat.*, May.
6. E. A. Hanushek (1974). 'Efficient estimators for regressing regression coefficients.' *The American Statistician*, May.
7. C. W. J. Granger, R. F. Engle, R. Ramanathan, and A. Andersen (1979). 'Residential load curves and time-of-day pricing: An econometric analysis.' *J. of Econometrics*, **9**, 13–32.
8. H. D. Burbank (1977). 'The Connecticut peak-load pricing experiment.' Anthony Lawrence (Ed.), *Forecasting and modelling time-of-day and seasonal electricity demands*, Electric Power Research Institute, EA 578-SR.
9. Electric Power Research Institute (1978). *Long-term residential load forecasting*, EA 584, Vols 1 and 2.
10. U.S. Department of Commerce (1959). *Notes on temperature–humidity index*, LS 5922.

Comparative Models for Electrical Load Forecasting
Edited by D. W. Bunn and E. D. Farmer

CHAPTER 10

Implementation of State Estimation and Data Validation in Electric Power Systems

M. R. Irving and M. J. H. Sterling

1 INTRODUCTION

Automatic or computer-assisted control of power generation, transmission, and distribution systems requires reliable data on the present network operating conditions. In general, a large number of transducers are available within the power system to measure such parameters as voltage, power flow, current, and tap position, the readings being telemetered to a remote computer control system. Data validation and state estimation processes are required to combine the measurements with a mathematical model of the network and with switch status information to generate an optimal estimate of the network state. If transient behaviour is neglected the state of a power system can be characterized by the voltage magnitude and phase angle at each network node. The optimal estimate should enable the calculation of more accurate values than those obtained from raw measurements and should provide estimates of quantities which are not physically measured. Telemetered values are subject to random noise and to gross errors associated with equipment malfunction. Capabilities for noise filtering and data validation (or bad data rejection) are therefore necessary features of state estimation techniques.

The present chapter discusses state estimation from a practical point of view and gives particular emphasis to implementational details and to the relationship between state estimation and other facets of power system control. The development of weighted least-squares and linear programming based methods which utilize sparse matrix techniques is described. Computational experience with real-time algorithms is presented and the application of state estimation to large-scale networks is considered.

2 STATE ESTIMATION AND DATA VALIDATION FUNCTIONS

State estimation and data validation functions must be combined with further processes to form an integrated system for electrical network control.[1] The major components of such a system are illustrated in Figure 10.1. The configuration of the overall scheme highlights similarity with conventional automatic controllers. Information is monitored via the telemetry system, load monitoring, data validation, and state estimation subsystems. Feedback control is effected via load frequency control, generation rescheduling, and load shedding, and feedforward control is implemented in the load prediction, security analysis, economic dispatch, and unit commitment phases.

Recent work[2–3] has indicated that data validation may be performed simultaneously at a number of different levels. This is shown in Figure 10.2 which expands the state estimation and data validation subsystem of Figure 10.1. Data obtained from individual substations or groups of substations may be processed by a series of parallel substation validation tasks. The resulting partially corrected measurements may be further processed by one or more area validation tasks. State estimation is then simplified by the

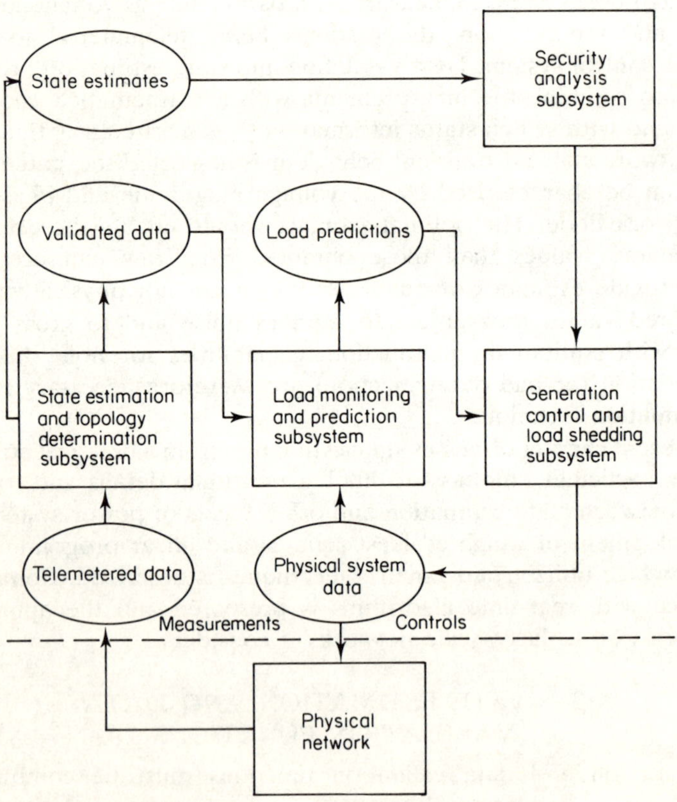

FIGURE 10.1 Overall software configuration

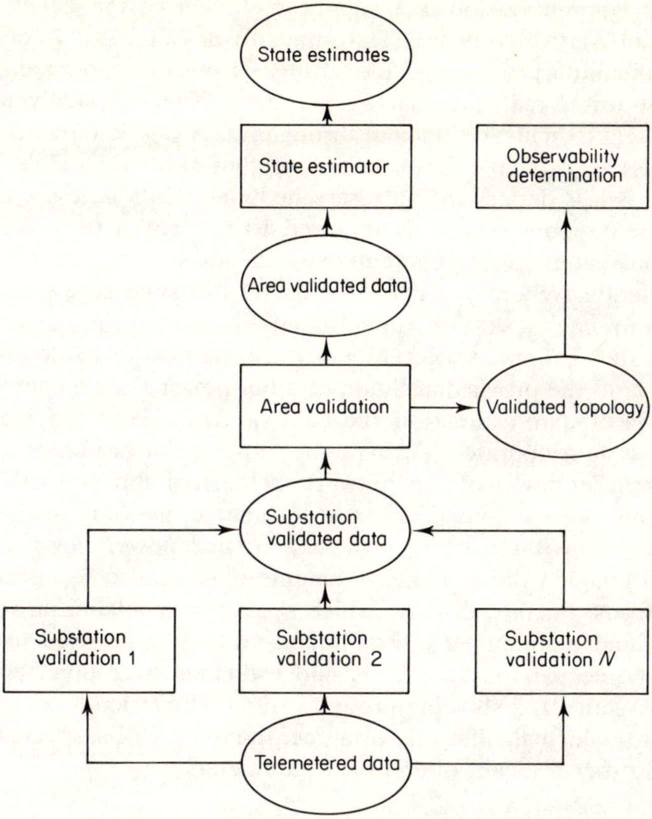

FIGURE 10.2 State estimation and data validation subsystem

elimination of a large proportion of the bad data at lower levels and by the creation of information on measurement reliability which may provide an iterative state estimator with a good starting position. However, the facility for bad data rejection at state estimation level remains important in order to diagnose gross errors which span substation or area boundaries.

The timing of task execution should be achieved as far as possible within the task itself. In this way every task or subsystem should be capable of being initiated or aborted at any time without compromising the integrity of the overall control system. Generally, state estimation and data validation tasks are run when one of the following criteria is satisfied:

(a) a run has been requested by the operator
(b) a time limit (say 10 mins) has expired since the last run
(c) a network connectivity change has been detected
(d) a significant change (say 5%) in any measurement has been registered.

Inter-task communication is a significant aspect of the development of power system control software. The input of measurement information to the state estimation process and the storage of optimal estimates imposes a requirement for a real-time database system. This is usually a memory resident system to achieve sufficient throughput. A shared data rather than a message passing principle is preferred so that the availability of shared information is less dependent on task scheduling. Data access synchronization via a semaphore system is required to prevent data inconsistencies. However, provision must be made to override any semaphores which remain set inadvertently. Where possible, it is useful to define access privileges to shared data for each task. For example, only the state estimator is permitted to write the state estimates whereas many tasks are permitted to read them.

In addition to the provision of data to other power system control tasks, a vital function of state estimation and data validation is to provide information to the system operator. Perhaps the most useful display is a 'one-line diagram' representation of the network or part of the network with raw measurements and validated or state estimated measurements superimposed. State estimates may be provided for line power flows and busbar voltages, although validated measurements must suffice for generator infeeds and loads. Further displays which should be available on request are trend plots and alphanumeric lists. Trend plots may be used to show the variation of selected measurements and estimates over time, and various lists may be generated showing present estimated overloads, measurements and estimates which significantly disagree, the history of suspected bad data points, and other relevant operational parameters.

3 PROBLEM FORMULATIONS

A wide variety of mathematical formulations have been proposed for state estimation and to a lesser extent data validation. The majority of these demand the solution of an overdetermined set of sparse linear equations

$$A\mathbf{x} = \mathbf{b} + \mathbf{e} \tag{10.1}$$

where

A is a matrix $(m \times n)$
$\mathbf{x}$ is a state vector $(n \times 1)$
$\mathbf{b}$ is a measurement vector $(m \times 1)$
$\mathbf{e}$ is an error vector $(m \times 1)$.

Numerical solution of this subproblem is the critical factor in the overall performance of the state estimator and is the key to efficient computation for large-scale networks. Equation (10.1) may be obtained by a Newton–Raphson linearization of the nonlinear measurement equations[4] or via the fast decoupled[5] or D.C. approximation methods. In the Newton–Raphson case an iterative outer loop is employed and A varies at each iteration. The fast decoupled approach is iterative but with a constant A matrix, and the D.C. method is non-iterative. The number of measurements is greater than the number of states to give some redundancy and matrix A has full rank, otherwise the network is defined as unobservable. It is possible to take advantage of the measurement redundancy in two ways. Firstly the effect on the final estimate $\mathbf{x}$ of small errors may be reduced, and secondly the measurements associated with large errors may be identified and their contribution to the solution reduced or eliminated.

There is an infinite number of solutions to equation (10.1) and therefore an objective function may be defined to select a particular solution which is optimal in some way. The weighted least-squares method minimizes

$$\mathbf{e}^{\mathrm{T}} W \mathbf{e} \tag{10.2}$$

where W is a weighting matrix.

The weighting matrix is chosen to reflect relative measurement accuracy and is usually taken as diagonal. This approach is particularly useful where the errors follow a normal distribution with known covariance. However, the method does not have any intrinsic bad data rejection properties. Data validation using weighted least-squares is usually accomplished via post processing schemes[6] which are often computationally expensive.

An alternative suggestion[7] is the minimization of the sum of weighted error modulus terms:

$$\mathbf{V}^{\mathrm{T}} |\mathbf{e}| \tag{10.3}$$

where

$\mathbf{V}$ is a weighting vector

$|\mathbf{e}|$ is a vector of error moduli.

This objective function does not place as much emphasis on large error terms as the least-squares function, and results in a more accurate estimate where bad data is present.

4 THE WEIGHTED LEAST-SQUARES APPROACH

The numerical solution of equation (10.1) subject to the weighted least-squares criterion (equation (10.2)) is often obtained by means of the normal equations:

$$A^{\mathrm{T}} W A \mathbf{x} = A^{\mathrm{T}} W \mathbf{b}. \tag{10.4}$$

Equation (10.4) may be obtained analytically and shows that the optimal estimate $\mathbf{x}$ can be determined by the solution of a set of n equations in n unknowns. The normal matrix $A^{\mathrm{T}}WA$ is symmetrical and sparse. However, there are a number of disadvantages associated with this technique. The normal matrix is less sparse than matrix A, and depending on the network and measurement topology the loss of sparsity can be very significant.

Equation (10.4) is not very well conditioned numerically. Also, the efficient construction of a suitably indexed sparse representation of $A^{T}WA$ is not simple to achieve.

An alternative technique for the solution of the weighted least-squares estimation problem is given by the augmented equations method.[8] The necessary conditions for a weighted least-squares solution may be written in matrix form as

$$\begin{bmatrix} 0 & I & A \\ I & W & 0 \\ A^{T} & 0 & 0 \end{bmatrix} \begin{bmatrix} \boldsymbol{\lambda} \\ -\mathbf{e} \\ \mathbf{x} \end{bmatrix} = \begin{bmatrix} \mathbf{b} \\ 0 \\ 0 \end{bmatrix} \qquad (10.5)$$

where

λ is a vector of lagrange multipliers
I is a unit matrix.

The augmented equations may be solved directly using a general sparse linear equation solution technique. Matrix A is duplicated although its original sparsity is now preserved. The computational burden imposed by sparse matrix factorization is approximately proportional to matrix order but varies according to the square of matrix density.[9] This lends support to the proposition that the augmented matrix method may be computationally superior to the normal matrix method in the power systems case. Also, it is relatively simple to construct the sparse representation of the augmented matrix effectively. Some computational results obtained with an augmented matrix algorithm are presented in a later section.

5 THE LINEAR-PROGRAMMING APPROACH

The solution of overdetermined linear equations (equation (10.1)) subject to a minimum error modulus criterion (equation (10.3)) may be found by computing the optimal feasible solution of an equivalent linear program[7]

$$\text{Min } \mathbf{V}^{T}\mathbf{e}_1 + \mathbf{V}^{T}\mathbf{e}_2 \qquad (10.6)$$

such that

$$A\mathbf{x} + \mathbf{e}_1 - \mathbf{e}_2 = \mathbf{b}$$

$$\mathbf{e}_1, \mathbf{e}_2 \geqslant 0.$$

The error vector **e** has been replaced by positive and negative components so that the moduli terms may be represented in the objective function. It can easily be shown that corresponding elements of $\mathbf{e}_1$ and $\mathbf{e}_2$ will not both be nonzero at the solution. Equation (10.6) may be represented in tableau form as:

0	$\mathbf{V}^{T}$	$\mathbf{V}^{T}$	
A	I	$-I$	$\mathbf{b}$

(10.7)

The linear program has a number of special features which may be utilized in deriving an efficient solution algorithm based on the sparse primal revised simplex method.[10]

(a) An optimal solution of the linear program will be determined by n of the original linear equations (the 'accepted' equations), while the remaining $m-n$ equations are said to have been rejected.
(b) The vector $\mathbf{x}$ is composed of free variables which are not restricted by the usual positivity requirements.
(c) The cost function is only nonzero for a subset of problem variables. This simplifies the computation of relative costs in the primal revised simplex method.
(d) Only a subset of the possible basic solutions are relevant; i.e. those which include all of the columns of A and some columns of I and $-I$ in the basis matrix. The columns which are selected from I and $-I$ mark those equations which are 'rejected'. The basis exchange process is simplified in that only columns of I and $-I$ need to be considered for exchange.
(e) A phase one simplex procedure to achieve feasibility is not necessary provided the user can supply an initial estimate of the set of equations which are 'accepted'. This feature is particularly useful if a previous solution to a related problem is available, as in the Newton-Raphson algorithm or when 'tracking' in a real-time system.
(f) The basis matrix is highly sparse and an efficient basis updating technique[11] may be applied.
(g) The unit matrices I and $-I$ may be stored implicitly to simplify the calculation of relative costs.
(h) The basis matrix retains a structure which enables factorization using a reduced basis method (see Appendix).

6 COMPUTATIONAL RESULTS

To illustrate the performance of the algorithms outlined in Sections 4 and 5, FORTRAN 77 implementations have been applied to the D.C. state estimation of the IEEE 30 node test system[12] with the measurement configuration shown in Figure 10.3. Computational results obtained using a 32-bit minicomputer (Perkin Elmer 3230) are presented in Table 10.1. In order to demonstrate the restarting capability of the linear programming method, an initial estimation run was followed by the further addition of approximately 1% random noise to the measurements and a repeat solution starting from the previously obtained set of accepted equations. It can be seen that the restarted estimation demanded considerably less computer time than the initial solution. The augmented matrix method is non-iterative and computation time therefore remains constant.

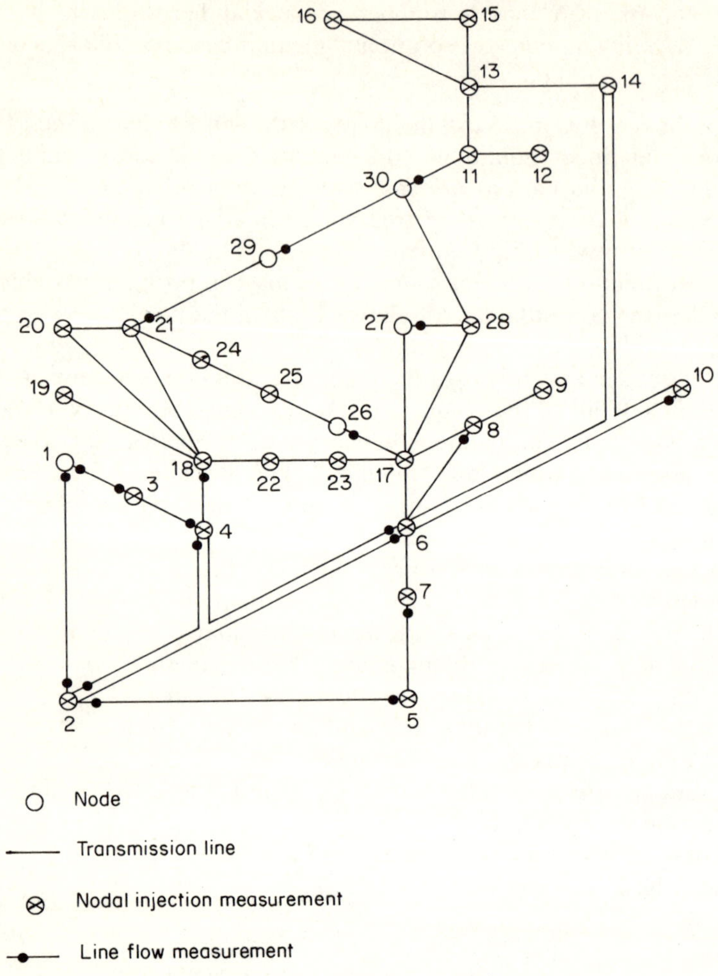

FIGURE 10.3 Measurement configuration for 30-node test system

Table 10.2 shows measured line flows together with state estimates obtained using the weighted least-squares method and the linear programming method, and Table 10.3 lists the measured and estimated nodal active power injections. The results show the effect of a sign reversal error in the measurement of active power flow in line 1. The presence of this gross error greatly perturbs the weighted least-squares state estimate leading to many large errors in neighbouring measurements. This effect is not apparent in the linear programming estimate which correctly concentrates the error on the bad data point.

TABLE 10.1 Computational requirements

IEEE 30 node system	No. of state variables	29
D.C. state estimation	No. of flow measurements	20
	No. of injection measurements	25
Weighted least-squares	Sum of squared errors	1.56
augmented matrix	Sum of error moduli	3.81
method	Solution time (sec)	0.58
Linear programming	Sum of squared errors	1.98
method	Sum of error moduli	1.52
initial solution	Solution time (sec)	1.56
	No. of iterations	27
Linear programming	Sum of squared errors	2.00
method	Sum of error moduli	1.53
repeat solution	Solution time (sec)	0.46
	No. of iterations	7

In addition to the results shown in the tables some experience has been obtained in A.C. state estimators and various data validators in a simulated real-time environment. As may be expected, weighted least-squares estimates have been found to be very favourable from the point of view of noise filtering. The storage and computer time requirements of the augmented matrix algorithm vary approximately linearly with network size, making the algorithm particularly attractive for application to large-scale systems. Further work is required to develop a suitable technique to enable bad data rejection in this method. It is hoped that a scheme based on iterative reweighting may prove suitable.

The linear programming approach has been found to be very efficient for data validation and bad data elimination. The computational efficiency of this method is greater for low-redundancy systems. Networks with high measurement redundancy may result in a large number of simplex iterations, as the number of vertices in the linear program model increases rapidly with the number of measurements. However, it has been observed that the majority of the useful work in the linear programming estimator occurs during the early iterations. Bad data is generally eliminated first and the remaining iterations simply consider various combinations of valid measurements to locate the optimal solution. Research is underway to define a reliable early termination criterion for the algorithm. Initial experience shows that the number of iterations needed may usually be halved. Experience with the reduced basis formulation indicates that a further saving in computer time may be realized. Although the storage requirement of the linear programming approach depends linearly on network size the solution time for a sparse matrix implementation varies approximately quadratically.

TABLE 10.2 State estimated line flows

Line no.	Node 1		Node 2	P_{meas} (p.u.)	P_{est} WLS (p.u.)	P_{error} WLS (p.u.)	P_{est} LP (p.u.)	P_{error} LP (p.u.)
1	1		2	−0.708*	0.231	−0.939	0.700	−1.408
		reverse		−0.700	−0.231	−0.469	−0.700	0.000
2	1		3	0.438	0.378	0.060	0.429	0.009
		reverse		−0.429	−0.378	−0.051	−0.429	0.000
3	2		4	0.317	0.349	−0.032	0.315	0.002
		reverse		−0.311	−0.349	0.038	−0.315	0.004
4	2		5	0.509	0.497	0.012	0.497	0.012
		reverse		−0.497	−0.497	0.000	−0.497	0.000
5	2		6		0.416		0.402	
		reverse		−0.397	−0.416	0.019	−0.402	0.005
6	3		4		0.103		0.404	
		reverse		−0.404	−0.103	−0.301	−0.404	0.000
7	4		6		0.306		0.389	
		reverse		−0.396	−0.306	−0.090	−0.389	−0.007
8	4		18		0.239		0.241	
		reverse		−0.241	−0.239	−0.002	−0.241	0.000
9	5		7		−0.044		−0.045	
		reverse		0.045	0.044	0.001	0.045	0.000
10	6		8		0.130		0.133	
		reverse		−0.133	−0.130	−0.003	−0.133	0.000
11	6		10		0.091		0.120	
		reverse		−0.119	−0.091	−0.028	−0.120	0.001
12	6		14		0.148		0.150	
13	6		17		0.114		0.116	
14	7		6		−0.245		−0.273	
15	8		17		0.330		0.333	
16	9		8		0.200		0.200	
17	10		14		0.025		0.020	
18	11		13		−0.040		−0.039	
19	12		11		−0.036		−0.035	
20	13		15		0.062		0.061	
21	14		13		0.173		0.169	
22	15		16		0.037		0.037	
23	16		13		−0.070		−0.069	
24	17		27		0.159		0.169	
25	17		28		0.078		0.080	
26	18		20		0.076		0.077	
27	18		21		0.171		0.176	
28	18		22		0.068		0.075	
29	19		18		0.199		0.200	
30	20		21		0.014		0.015	
31	21		24		0.058		0.066	
32	21		29	0.055	0.053	0.002	0.055	0.000
33	22		23		0.032		0.040	
34	23		17		−0.059		−0.050	
35	24		25		0.027		0.034	

TABLE 10.2—(*continued*)

Line no.	Node 1	Node 2	P_{meas} (p.u.)	P_{est} WLS (p.u.)	P_{error} WLS (p.u.)	P_{est} LP (p.u.)	P_{error} LP (p.u.)
36	25	26		−0.067		−0.061	
37	26	17	−0.092	−0.092	0.000	−0.092	0.000
38	27	28	−0.013	−0.010	−0.003	−0.027	0.014
39	28	30		0.068		0.053	
40	29	30	0.022	0.021	0.001	0.022	0.000
41	30	11	−0.002	−0.004	0.002	−0.004	0.002

* Bad data point represented by the reversal of the sign of the true measurement of 0.708.

TABLE 10.3 State estimated nodal injections

Node no.	P_{meas} (p.u.)	P_{est} WLS (p.u.)	P_{error} WLS (p.u.)	P_{est} LP (p.u.)	P_{error} LP (p.u.)
1		0.609		1.129	
2	0.533	1.031	−0.498	0.514	0.019
3	−0.024	−0.275	0.251	−0.025	0.001
4	−0.076	0.093	−0.169	−0.089	0.013
5	−0.542	−0.541	−0.001	−0.542	0.000
6	0.000	0.007	−0.007	0.000	0.000
7	−0.228	−0.202	−0.026	−0.228	0.000
8	0.000	0.000	0.000	0.000	0.000
9	0.200	0.200	0.000	0.200	0.000
10	−0.100	−0.066	−0.034	−0.100	0.000
11	0.000	0.000	0.000	0.000	0.000
12	−0.035	−0.036	0.001	−0.035	0.000
13	0.000	0.000	0.000	0.000	0.000
14	0.000	0.000	0.000	0.000	0.000
15	−0.024	−0.025	0.001	−0.024	0.000
16	−0.106	−0.107	0.001	−0.106	0.000
17	−0.058	−0.056	−0.002	−0.058	0.000
18	−0.112	−0.123	0.011	−0.112	0.000
19	0.200	0.199	0.001	0.200	0.000
20	−0.062	−0.062	0.000	−0.062	0.000
21	−0.082	−0.074	−0.008	−0.070	−0.012
22	−0.035	−0.036	0.001	−0.035	0.000
23	−0.090	−0.091	0.001	−0.090	0.000
24	−0.032	−0.031	−0.001	−0.032	0.000
25	−0.095	−0.094	−0.001	−0.095	0.000
26		−0.025		−0.031	
27		−0.169		−0.196	
28	0.000	0.000		0.000	
29		−0.032		−0.033	
30		−0.093		−0.079	

However, it is believed that the application of the restarting technique in a tracking environment will allow solution times to be effectively constrained.

7 CONCLUSIONS

The primary functions of modern power system control centres are security enhancement, energy optimization, and plant monitoring. Within this framework a state estimation and data validation facility is normally a prerequisite since it enables available measurements to be processed to obtain network data which is as complete and consistent as possible. The elimination of erroneous information, the reduction of the effects of noise by filtering, and the replacement of lost or unavailable measurements are consequently essential features of any on-line monitoring and control system.

The present chapter has concentrated on some of the aspects of state estimation and data validation which are important for practical implementation in computer control centres. Methods for minimum error modulus estimation using linear programming and weighted least-squares estimation using the augmented matrix algorithm have been described, together with details of computational techniques which enable efficient implementation for medium- and large-scale problems. The computational requirements which have been presented indicate that the algorithms are suitable for application in real time to large-scale electrical power networks using inexpensive minicomputer hardware.

REFERENCES

1. M. J. H. Sterling and M. R. Irving (1985). 'Integrated software for power system control.' In M. Singh (Ed.), *Encyclopaedia of systems and control*, Pergamon Press.
2. M. R. Irving and M. J. H. Sterling (1982). 'Substation data validation.' *Proc. IEE*, C, **129,** 119–122.
3. B. C. Clewer, M. R. Irving, and M. J. H. Sterling: 'Linear programming state estimation in power systems.'
4. F. C. Schweppe and E. J. Handschin (1974). 'Static state estimation in electric power systems.' *Proc. IEEE*, **62,** 972–982.
5. B. Stott and O. Alsac (1974). 'Fast decoupled loadflow.' *IEEE Trans.*, **PAS. 93,** 859–869.
6. F. Brousolle (1978). 'State estimation in power systems: Detecting bad data through the sparse inverse matrix method.' *IEEE Trans.*, **PAS. 97,** 3, 678–682.
7. M. R. Irving, R. C. Owen, and M. J. H. Sterling (1978). 'Power system state estimation using linear programming.' *Proc. IEE*, **125,** 9, 879–885.
8. I. S. Duff (1977). 'A survey of sparse matrix research.' *Proc. IEEE*, **65,** 4, 500–535.
9. F. C. Aschmoneit, N. M. Peterson, and E. C. Adrian (1977). 'State estimation with equality constraints.' *Power Industries Computer Applications Conference. 1977.*

10. D. G. Luenberger (1973). *Introduction to linear and nonlinear programming*, Addison-Wesley.
11. J. K. Reid (1975). '*A sparsity-exploiting variant of the Bartels–Golub decomposition for linear programming basis*,' Report CSS 20, AERE Harwell.
12. L. L. Freris and A. M. Sasson (1968). 'Investigation of the loadflow problem.' *Proc. IEE*, **115**, 10, 1459–1470.

APPENDIX THE REDUCED BASIS FORMULATION

The linear program model presented in Section 5 results, at every iteration, in an $m \times m$ basis matrix of the form:

$$[A \,\vdots\, E] \tag{A10.1}$$

where E is composed of certain columns of I and $-I$. If the rows of the basis matrix are reordered and negated where necessary, the following structure may be obtained:

$$\left[\begin{array}{c:c} Aa & 0 \\ \hdashline Ar & I \end{array}\right]. \tag{A10.2}$$

The necessary calculations for the revised simplex method may now be performed by factorizing and updating the $n \times n$ matrix Aa only, and using the following expressions:

$$\left[\begin{array}{c:c} Aa & 0 \\ \hdashline Ar & I \end{array}\right]^{-1} = \left[\begin{array}{c:c} Aa^{-1} & 0 \\ \hdashline -ArAa^{-1} & I \end{array}\right] \tag{A10.3}$$

$$\left[\begin{array}{c:c} Aa & 0 \\ \hdashline Ar & I \end{array}\right]^{-\mathrm{T}} = \left[\begin{array}{c:c} Aa^{-\mathrm{T}} & -Aa^{-\mathrm{T}}Ar^{\mathrm{T}} \\ \hdashline 0 & I \end{array}\right]. \tag{A10.4}$$

This approach leads to a reduced basis matrix Aa which is factorized and updated more rapidly at the expense of an increased overhead incurred by equations (A10.3) and (A10.4).

Comparative Models for Electrical Load Forecasting
Edited by D. W. Bunn and E. D. Farmer

CHAPTER 11

Short-term Forecasting at British Gas

J. L. Piggott

1 INTRODUCTION

Natural Gas is supplied to the British Gas Corporation via onshore terminals of the suppliers. The gas is transmitted through a high-pressure network of pipes to the 12 regions of the Corporation. The gas moves relatively slowly in the pipes, and to ensure that customers are supplied at all times accurate forecasts of the demand for gas are required. These forecasts cover 3 different periods. A daily forecast is required so that plant such as compressor stations, LNG storage, and low-pressure gas holders can be operated economically and so as to maintain a secure supply. Hourly forecasts are needed to ensure that supply and demand are matched and are used to monitor the system. Looking very much further ahead weekly forecasts, up to the end of the contract year, are needed so that any contractual and transmission constraints can be taken into account when scheduling the gas supplies from each of the sources.

The main use for the gas is space heating and therefore the prevailing weather conditions have a large effect on the amount of gas required by customers. The demand in a region follows a diurnal pattern with a peak occurring in the early evening and a trough in the early hours of the morning. This diurnal pattern is mainly smoothed out in the regions by the use of storage, and the demand on the national system has a much flatter profile. At the beginning of each gas day, regions adjust their demands to take into account the next day's demand, and this is seen as a change in demand level on the national system. The demand also follows a weekly pattern with a trough at the weekend, and a peak in the middle of the week. This weekly pattern is sometimes difficult to see due to the effect the weather has on the demand. There is also an annual pattern, with low points in August and high points in January and February.

In order to produce accurate demand forecasts it is necessary to have available good forecasts of the weather conditions throughout the UK. This chapter describes the computerized system developed at London Research

Station for decoding wather forecasts sent via the British Telecom telex system. The report also describes how these forecasts are sorted, then displayed, and finally used to produce the three types of gas demand forecast required by the Grid Control Engineers.

2 THE WEATHER FORECASTS COLLECTION SYSTEM

2.1 History

During the late 1960s and early 1970s most of the gas supplied to customers was manufactured locally. At that time each regional gas board received weather forecasts from their local weather centre via the telex system. These forecasts were used to forecast the demand for gas and contained a variety of information, depending on the region and weather centre. After the discovery of natural gas and the subsequent development of the network of pipes which carries the gas to the regions, it was found that it was necessary to forecast the gas demand for the whole country, rather than for particular localities. Since all the forecasts contained different information, and arrived at different time intervals, it was very difficult to get an accurate picture of the weather throughout the country. It was therefore decided that if possible a set format should be adopted for weather forecasts. Discussions between London Weather Centre, the London area regions, and BGC Headquarters led to a set format being tried. It was hoped that if successful other regions and weather centres would adopt the same format, and join the so-called METGAS system.

2.2 Timing and contents of forecasts

As mentioned previously, different regions received weather forecasts at different time intervals, varying from as often as three-hourly to as rarely as once per day. When deciding how often to take forecasts, the main decision that has to be made is how often forecasts must be received in order that any changes in weather conditions can be picked up as quickly as possible. The factor working against this is the cost of taking forecasts at very short intervals. Since the highest frequency taken by a region was eight per day and the lowest one per day, it was thought sensible to opt for somewhere between these extremes, and six per day was finally adopted. The first forecast is received at 4 a.m., two hours before the start of the 'gas day', and the others at 0800, 1200, 1530, 2000, and 2400. The forecast at 1600 is received half an hour earlier, to help both headquarters and regions to forecast demand for the following day, before the end of the working day.

The information contained in the local weather forecasts sent to individual regions varied considerably. It was not feasible to include all information required by every region and it was therefore necessary to limit the amount of information contained in each forecast to a few key variables. The demand for gas is mainly dependent on the outside temperature, but is also affected by the wind, sunshine, rainfall, humidity, and other associated phenomena. Of those mentioned only the wind has a large quantifiable effect. Each forecast therefore contains information about temperature and wind (both speed and direction) for the next 24 hours. Information about other effects is updated at 0800 and 1530, and is sent in the form of comments. In addition forecasts of maximum and minimum temperatures for the next 3 days are also only updated at 0800 and 1530. A more detailed description of the actual telex format is given in Section 2.4.

2.3 The development of the computerized system

As more weather centres joined the system (currently there are eight) it was soon realized that it was a time-consuming task composing the individual forecasts received via the telex system. Having a similar format for all forecasts meant that it was possible to decode the telexes using a computer. A program was therefore written to decode the telexes and store the forecast in an on-line database. Initially the telexes were received punched onto computer cards, and read into the computer. This was soon superseded by a system that could read the telex papertapes directly. Such a system still exists and is used only as a back-up facility in case of emergency.

British Gas Headquarters have two UNIVAC 1100 series computers, which are used for grid control. One is used as the main or primary machine, the other as a back-up. The two machines are permanently linked and information is transferred from one machine to the other, so that in the event of main machine failure the back-up machine can take over. Each machine has a real-time telex handler which controls the input and output of telex messages on the telex circuits, using the direct dialling facilities. There are two telex lines associated with each machine. The current METGAS system involves the use of these direct telex facilities, thus the use of telex tapes and hence manual intervention is no longer required.

The weather centres have five telex numbers which they can use to send weather forecasts to the computer system: two connected to the main computer, two to the back-up, and one connected to the manual paper-tape system. Under normal circumstances the first two numbers are used, but if the lines are absent or occupied, then one of the second pair of numbers is used, and only if all else fails the last number is tried. When a weather centre dials the computer, it checks the answer-back code of the sender, and if recognized creates a computer file, into which the telex can be put. The computer then sends a 'ready' message to the sender indicating the forecast can be sent. On receiving the end-of-message character, the line is closed down, and a run within the computer is started up to process the telex.

2.4 Decoding a METAGAS telex

The standard telex format is shown in Table 11.1. Each set of information is preceded by a code letter A to O, and starts on a new line. The first line must be the station identifier, sections A to G are compulsory at all times, and the last line must be ENDS. A typical telex is shown in Figure 11.1. When decoding a telex, checks have to be made on the authenticity of the data, to prevent as far as is possible the storing of mistyped or incorrect data. The decoding program can be run in two modes, normal or on-line. In on-line mode the program will correct errors in the date to produce today's date, and if there is an error in the time section, and the actual time is within one hour of a forecast time, then the time is corrected. In normal mode the date and time must be correct, otherwise processing will stop. The normal mode is used for inputting data at a later date.

The forecast data itself goes through several checks to ensure that as few errors as possible get stored in the database. The temperature data is checked first and upper and lower limits are put on the temperatures. The

TABLE 11.1 Standard METGAS telex format

Section	Contents	Forecasts
A	Station mnemonic	All
B	Date of forecast	All
C	Time of forecast	All
D	Actual temperatures for last 2 two hour periods (to nearest °C)	All
E	Two hourly forecast temperatures for next 24 hours (at least)	All
F	Actual wind speed followed by wind direction	All
G	At least 6 forecast wind speeds plus 6 directions	All
H	Forecast maximum for day 1	8+16
I	Forecast minimum for night 1	8+16
J	Forecast maximum for day 2	8+16
K	Forecast minimum for night 2 and Forecast maximum for day 3	16
L	Comments for today	8+16
M	Comments on temperature for next 24 hours	8+16
N	Comments on wind for next 24 hours	8+16
O	Outlook	8+16

```
>
>    1.       A LWC
>    2.       FROM LONDON WEATHER CENTRE
>    3.       3/11/81
>    4.       A       LWC
>    5.       B       3/11/81
>    6.       D       0800
>    7.       D       13     13
>    8.       E       14     14     15     15     15     14
>    9.       14      13     13     12     12     12
>   10.       F       08 SW
>   11.       G       12     14     16     16     15     10
>   12.       SW      SW     SW     SW     SW     WSW
>   13.       H       14
>   14.       I       08
>   15.       J       11
>   16.       L       MAINLY CLOUDY, DRY WITH BRIGHT INTERVALS TODAY.
>   17.               RAIN LATER IN THE NIGHT
>   18.       M       TENDENCY FOR FALL IN TEMPERATURE OVER NEXT FEW
>   19.               DAYS. TIMING UNCERTAIN
>   20.       N       WEST OR WESTNORTHWEST MODERATE TO FRESH
>   21.       O       CLOUDY WITH RAIN EARLY BECOMING DRY LATER WITH
>   22.               SUNNY OR CLEAR INTERVALS.
>   23.       TOD 0739
>   24.       24644  WEALWC G
>END DATA. ERRORS: NONE. TIME:  1.141 SEC. IMAGE COUNT: 24
>
```

FIGURE 11.1 A typical METGAS telex

maximum and minimum allowable temperatures are 40 °C and −20 °C; these limits trap most cases where temperatures are run together, e.g. 89 10 11 instead of 8 9 10 11. In such cases the program would check to see if there was too little data in the section, and if this were the case the 8 9 would be split. In some cases however this type of error may get through the system, e.g. 11 1 instead of 1 1 1. The expected amount of information per section for each forecast time is shown in Table 11.2. If there is an incorrect amount of information nothing is stored, but processing is continued.

Section F contains the actual windspeed in knots, followed by the wind direction to the nearest sixteenth (i.e. N, NNE, NE, ENE, etc.). Each direction may be written in full or abbreviated, except that the direction EAST must be written in full to avoid confusion with the error code E. Similarly Section G should contain the correct number of windspeeds followed by the directions; again there are checks on the data. The windspeed

TABLE 11.2 The expected amount of data per section at each forecast time

	Section							
Time	D	E	F	G	H	I	J	K
0800	2	12	1	6	1	1	1	—
1200	2	12	1	6	—	—	—	—
1530	2	19	1	9	1	1	1	2
2000	2	17	1	8	—	—	—	—
2400	2	15	1	7	—	—	—	—
2400	2	13	1	6	—	—	—	—

must be greater than or equal to zero, and be less than 50 knots, and there must be the correct amount of information in the forecast. Any undecoded wind directions are set to 'XXX'.

Sections H onwards are only received at 0800 and 1530. There are fewer checks on the maximum and minimum temperature forecasts than for the two-hourly temperatures, and no checks at all are made on the comments sections. The comments are in fact only stored for 24 hours, whereas all the other data is stored permanently. Should a forecast not have been received by 20 minutes after the forecast time, or the telex not decoded correctly, then the computer automatically sends the weather centre a telex requesting them to re-send their forecast. There are 48 forecasts received each day, and on average one or two need figures correcting by hand, and one or less are completely missing.

2.5 Display of the METGAS forecasts

Currently there are 8 weather centres sending weather forecasts to the computer system. The centres are distributed throughout the UK and their locations are shown in Figure 11.2. In order to compare weather conditions across the country a set of computer programs has been written to display the weather forecasts for the grid control engineers. The information is displayed by three programs, in the form of four tables, examples of which are shown in Figures 11.3(a)–(d). Table A7 (Figure 11.3a) shows the two-hourly forecast temperatures for the next 24 hours, the actual temperatures 1 and 3 hours ago, and the maximum and minimum forecast temperature for the next three days where available. A similar table follows showing the windspeed and directions. These tables enable the grid control engineers to see variations in weather across the country easily. It is also possible to list out the latest comments from each station, examples of which are shown in Figure 11.3(c), and finally the current temperatures can be displayed in map form as shown in Figure 11.3(d). These programs are used up to 6 times daily by the grid control engineers, as are the demand forecasting programs described in the next section.

3 DATA USED FOR MODELLING AND FORECASTING

3.1 Weather data

The only data available for modelling purposes is the 2-hourly forecast and actual temperatures, and the 4-hourly actual and forecast windspeeds and directions for the next 24 hours and the forecast maximum and minimum temperatures for the following 2 or 3 days. It is generally thought that

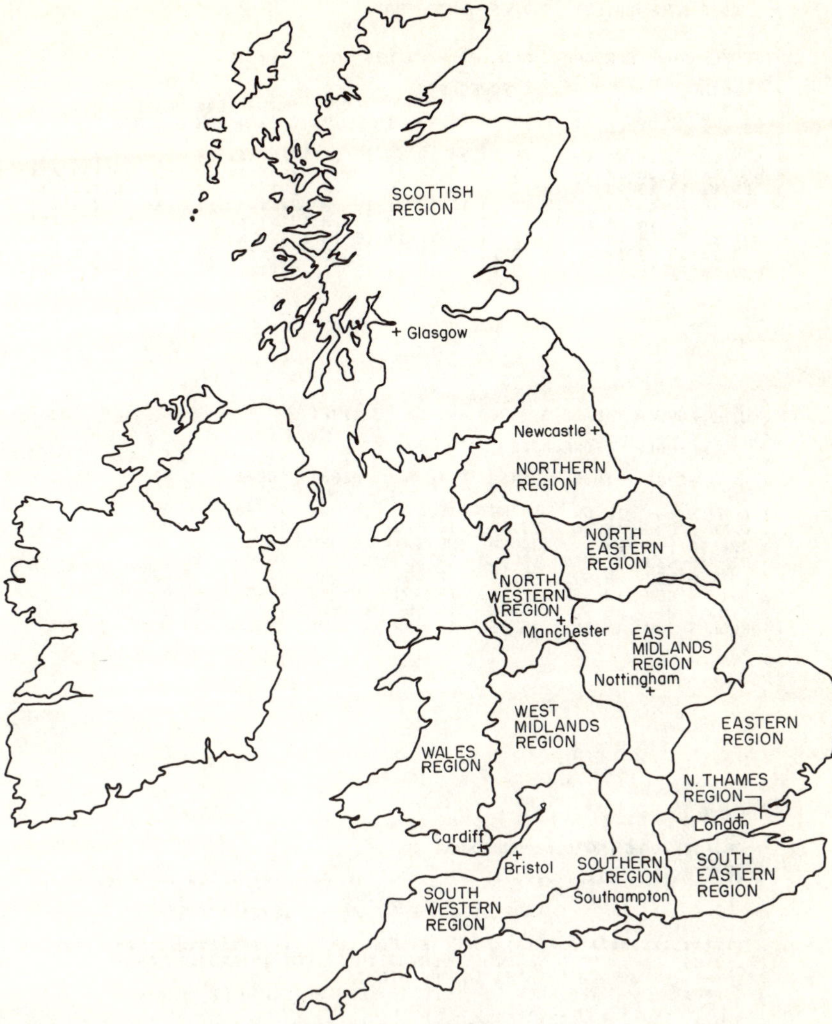

FIGURE 11.2 The eight METGAS stations from which weather forecasts are received every four hours

nationally any effect of wind direction is taken into account by the temperature (i.e. a lower temperature when a north-easterly is blowing), therefore only temperature and windspeed are considered. These data must be combined into 2-hourly, daily, and weekly national averages. To calculate the national temperature at any particular time, a weighted average of the eight individual station temperatures is taken. These weights are shown in Table 11.3 and represent the relative quantity of gas sold around each station.

```
>  TABLE A7 - LATEST WEATHER CENTRE DATA
>
>
> TEMPERATURE FORECASTS MADE AT 0800 HRS ON  3/11/81
>
>   ACTUALS                      FORECASTS                           4/11    5/11   6/11
>                                                                DAY NGT  DAY NGT DAY
> MET   5   7   9 11 13 15 17 19 21 23   1   3   5   7   9      MAX MIN  MAX MIN MAX
>
> LON 13 13 14 14 15 15 15 14 14 13 13 12 12 12****** 14    8   11 *** ***
> WAT 12 10 12 14 16 16 14 13 12 12 11   9   8   7****** 10    3   12 *** ***
> ABB 13 13 13 13 12 12 10 10   9   9   9   8   8   8******   8    3    8 *** ***
> RIN 12 12 13 14 15 15 14 13 13 12 11 10   9   8****** 11    4    8 *** ***
> SOT 13 13 13 14 15 15 14 14 13 13 13 13 13 13****** 13    7   11 *** ***
> BRI 13 12 14 15 15 15 15 14 13 12 11 10 10 10****** 12    5    9 *** ***
> TYN 13 13 14 14 14 14 13 12 11 10   9   7   5   5****** 10    2    9 *** ***
> RHO 11 11 12 13 14 14 13 12 12 12 11 11   9   9****** 12    3   10 *** ***
```

(a)

```
>
> WIND FORECAST MADE AT 0800 HRS ON  3/11/81
>
>     ACTUALS   FORECASTS
>
>      0600     1000     1400     1800     2200     0200     0600
>
> LON   8 SW    12 SW    14 SW    16 SW    16 SW    15 SW    10 WSW
> WAT   8 SW    11 SW    14 SW    16 SSW   12 W     10 WSW    9 NW
> ABB  19 S     20 S     20 SW    25 SW    20 WSW   20 WSW   15 WSW
> RIN  13 SSW   16 SSW   18 SSW   22 SSW   18 WSW   16 W     15 W
> SOT   3 WSW    5 SW     7 SW     8 SW     7 WSW    7 WSW    7 WSW
> BRI   6 SSW   12 SSW   17 SSW   18 SSW   18 SSW   15 W     15 NW
> TYN   0 N     15 SW    18 SW    18 SW    15 SW    15 W     15 W
> RHO   8 SSW   11 SW    15 SSW   17 SSW   17 SW    17 SSW   10 WNW
>
>
>
>
>
```

(b)

```
>
>FACILITY WARNING  100000000000
>READY
> ==========================================================
>
>                         LATEST COMMENTS ON   3 11 81 AT  800
>
> ************************FORECAST FOR LWC    *************************
> L:0+24          MAINLY CLOUDY  DRY WITH BRIGHT INTERVALS TODAY
>                 RAIN LATER IN THE NIGHT
> M:T+24          TENDENCY FOR FALL IN TEMPERATURE OVER NEXT FEW
>                 DAYS  TIMING UNCERTAIN
> N:W+24          WEST OR WESTNORTHWEST MODERATE TO FRESH
> 0:0>24          CLOUDY WITH RAIN EARLY BECOMING DRY LATER WITH
>                 SUNNY OR CLEAR INTERVALS
>
> ************************FORECAST FOR WAT    *************************
> L:0+24      VERY MILD SW LY FLOW WILL GIVE WAY TO NOTICEABLY COLDER
>          WEATHER OVERNIGHT
> M:T+24      WELL ABOVE NORMAL TODAY
> N:W+24      SW INCREASING TO STRONG LATER TODAY
> 0:0>24      NOTICEABLY COLDER  SCATTERED SHOWERS AT FIRST  THEN DRY
>          NIGHT FROST
>
> ************************FORECAST FOR ABB    *************************
> L:0+24      MOSTLY CLOUDY  OCCASIONAL RAIN  CONTINUOUS FOR A TIME THIS
>             AFTERNOON
> M:T+24      RATHER WARM BECOMING NORMAL BY TOMORROW
> N:W+24      STRONG W BECOMING MODERATE NW
> 0:0>24      SUNNY INTERVALS  SCATTERED SHOWERS
>          ?
```

(c)

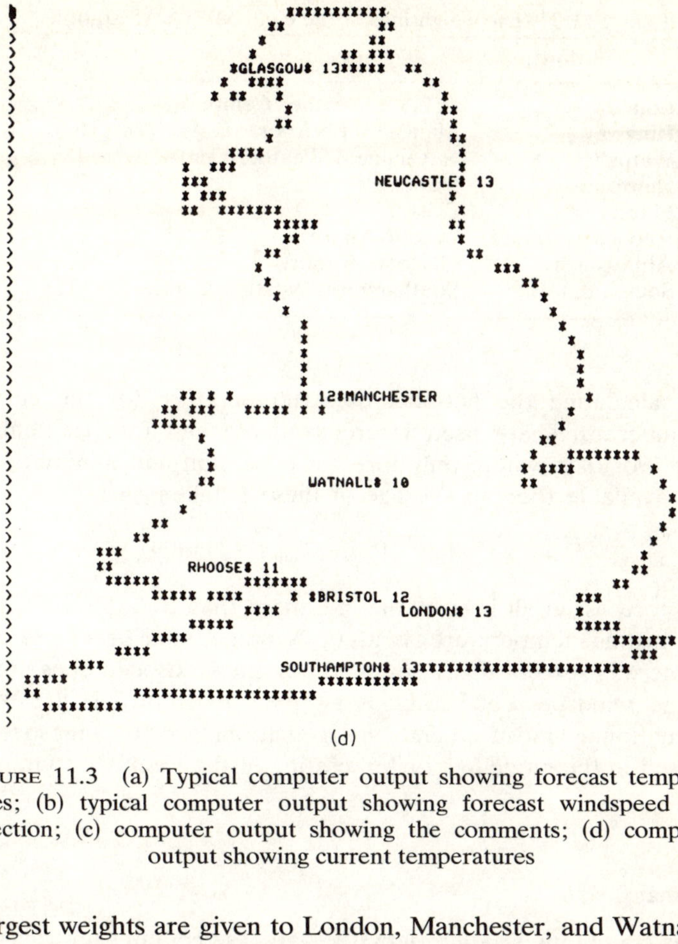

(d)

FIGURE 11.3 (a) Typical computer output showing forecast temperatures; (b) typical computer output showing forecast windspeed and direction; (c) computer output showing the comments; (d) computer output showing current temperatures

The largest weights are given to London, Manchester, and Watnall, being closest to the largest centre of population. The national average daily temperature is a straight average of the 12 two-hourly temperatures, and the national weekly temperature is the average of the 7 daily temperatures

$$T_h = \sum_{i=1}^{8} W_i t_i \qquad T_o = \sum_{h=1}^{24} T_h/24 \qquad T_w = \sum_{o=1}^{7} T_o/7$$

where

W_i = station weight

t_i = 2-hourly temperatures for station i

T_h = national hourly temperatures

T_o = national daily temperatures

T_w = weekly temperatures.

TABLE 11.3 The weightings of the eight METGAS stations

Station	Location	Weight (%)
London	London Weather Centre	27.4
Ringway	Manchester Airport	16.3
Watnall	Nottingham Weather Centre	11.1
Newcastle	Newcastle	7.6
Filton	Bristol	10.5
Rhoose	Cardiff Airport	6.2
Abbotsinch	Glasgow Airport	9.6
Southampton	Southampton Weather Centre	11.3

When calculating the national daily temperature for the current day, actual temperatures are used where available and forecast temperatures otherwise. For days where only forecast maximum and minimum temperatures are available then an average of these is taken, i.e.

$$T_o = (T_{\mathrm{max}} + T_{\mathrm{min}}) \div 2.$$

When no forecast at all is available, i.e. more than 3 days ahead, then SNT (seasonal normal temperature) is used. Average windspeeds are calculated in the same way, except that a seasonal normal windspeed does not exist, so an 'average' windspeed of 8 knots is assumed for all days in the future. The 4-hourly national windspeed and average national weekly windspeed are not usually used in the modelling or forecasting of the weekly or within-day gas demand.

3.2 Demand data

As well as containing weather data, the database contains information about gas demand. This is in the form of the total amount of gas taken out of the system each day, and the percentage of the total taken in each 2-hour period. The weekly demand is the sum of the daily demands. During periods of high demand industrial customers may have contracts which allow their supply to be interrupted, thus reducing the demand for gas. Since the unrestricted demand must be used for forecasting future demand, the amount of interruption is stored in the database, and subsequently added to the daily 'take' from the system to form the unrestricted demand. It is this unrestricted demand that is used as the basic data for both daily and weekly modelling. For 2-hourly demand modelling the actual 'take' figures are used because of the difficulties involved in estimating the amount of interruption on an hourly basis.

3.3 Modifications to the basic data

Before any modelling can take place, the raw data series must be converted to more suitable forms. These forms vary depending on whether wind is to be included, whether the data being modelled is hourly, daily, or weekly, or indeed on the time of year being considered. Each data set (i.e. hourly, daily, and weekly) will be considered separately.

3.3.1 Daily data

There are three modifications made to the daily data before modelling, one affecting the temperatures and two affecting the demand. The first involves modifying the temperatures above 14 °C to account for the decrease in response of the demand series to similar changes in temperatures below 14 °C. The modification takes the form of a transformation;

$$T_t = 13 + \left(\frac{1 - 0.8^{T-13}}{0.2}\right) \qquad T \geqslant 14.05\,°\mathrm{C}$$

$$T_t = T \qquad T < 14.05\,°\mathrm{C}$$

The function has a maximum value of 18 °C.

The second modification made to the raw data is to take account of holiday periods. The demand for gas decreases around all bank holiday periods, especially around Christmas and the New Year. For most holdays the percentage decrease is similar from year to year and can easily be estimated. The holiday which is most difficult is Christmas/New Year, which occurs on a different day of the week each year and consequently an estimate of the decrease must be made using past experience. The current holiday factors are shown in Table 11.4, the demand on each of these days being multiplied by the holiday factor before modelling commences.

The final correction made to the daily data takes account of the change to and from British Summer Time. In March and October when the clocks are put forward and backward 1 hour respectively, a 23-hour day and a 25-hour

TABLE 11.4 The daily holiday correction factors for 1981

	Wed	Thu	Fri	Sat	Sun	Mon	Tue	Wed
New Year	1.15	1.20	1.05	1.00	1.00	1.00	1.00	1.00
Easter	1.04	1.07	1.13	1.07	1.12	1.18	1.05	1.02
Mayday	1.00	1.00	1.00	1.10	1.10	1.15	1.00	1.00
Spring	1.00	1.00	1.03	1.05	1.10	1.30	1.25	1.15
Summer	1.00	1.03	1.05	1.00	1.04	1.19	1.07	1.03
Xmas	1.10	1.20	1.35	1.20	1.20	1.35	1.15	1.15

day are created. A pro-rate correction is made to the demand series, to stop these days distorting the data. Such a correction is thought to be adequate, due to small variation in demand on the national system from hour to hour.

3.3.2 Two-hourly data

Only one correction is made to the two-hourly data; this takes account of the change to British Summer Time, when a two-hour period is shrunk to one hour in the spring and expanded to 3 hours in the autumn. Again the correction is done on a pro-rata basis. No account is taken of holiday periods.

3.3.3 Weekly data

It was initially thought that the weekly data need only be corrected for the Easter holday since all the others occur at the same time each year. However, all the one-day holidays do move to the adjacent week every 5 or 6 years, due to there not being an exact number of weeks in a year. To account for this a correction factor is required for the week either side of a bank holiday. These factors are shown in Table 11.5.

It has been found unnecessary to correct the temperature series for temperatures above 14 °C. It is thought that this is due to the yearly factor in the data; no correction is applied to take account of the 23 and 25 hour days, since any correction would be less than 1% of the weekly demand.

4 MODELLING AND FORECASTING

4.1 Method

The technique used for modelling and forecasting all three data series is the Box–Jenkins method. A computer program written at LRS has been used to

TABLE 11.5 Weekly holiday correction factors for 1981

	Preceding week	Following week
Easter	1.04	1.05
Mayday	1.01	1.03
Spring	1.01	1.11
Summer	1.01	1.05
Xmas/New Year	1.12 (Xmas week)	1.16 (New Year week)

do all the analysis referred to in this section. The development of the modelling process is described briefly, together with an account of the forecasting performance of the various models. Each data set (daily, two-hourly, and weekly) is dealt with separately.

4.2 Modelling daily data

The computer program for analysis time series only allows one output series (demand) and one input series (temperature) to be analysed. Such a limitation means that to include a third series such as the windspeed, we must either correct for wind, or combine wind and temperature into a chill factor. Models have been developed using (i) temperature alone, (ii) the demand series corrected for windspeed, an estimate has to be made of the effect of windspeed on demand. The correction used in this analysis is 1% increase in demand for every 2.5 knots increase in windspeed between 8 and 15 knots. Hence no correction below 8 knots, and a 2.8% increase for windspeeds 15 knots and above. This simple wind correction has two major disadvantages: the sharp cut-offs at 8 and 15 knots; and the effect of a similar windspeed is the same at all temperatures. One possible way of accounting for such effects is to combine the temperature and windspeed into a 'chill factor'.

The general form of the 'chill factor' used in this work is

$$\text{CHILL} = \text{TEMPERATURE} - (T_{\max} - \text{TEMPERATURE}) \times F(\text{WIND})$$

where $T_{\max}$ is a parameter and $F(WIND)$ is a function of windspeed whose graph is similar to:

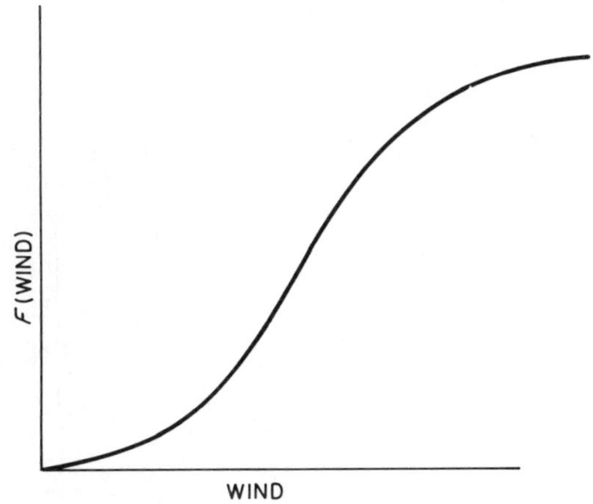

The ($T_{\max}$ − TEMPERATURE) term is included so that the chilling effect of the wind is greater at lower temperatures, and $T_{\max}$ is taken to be about

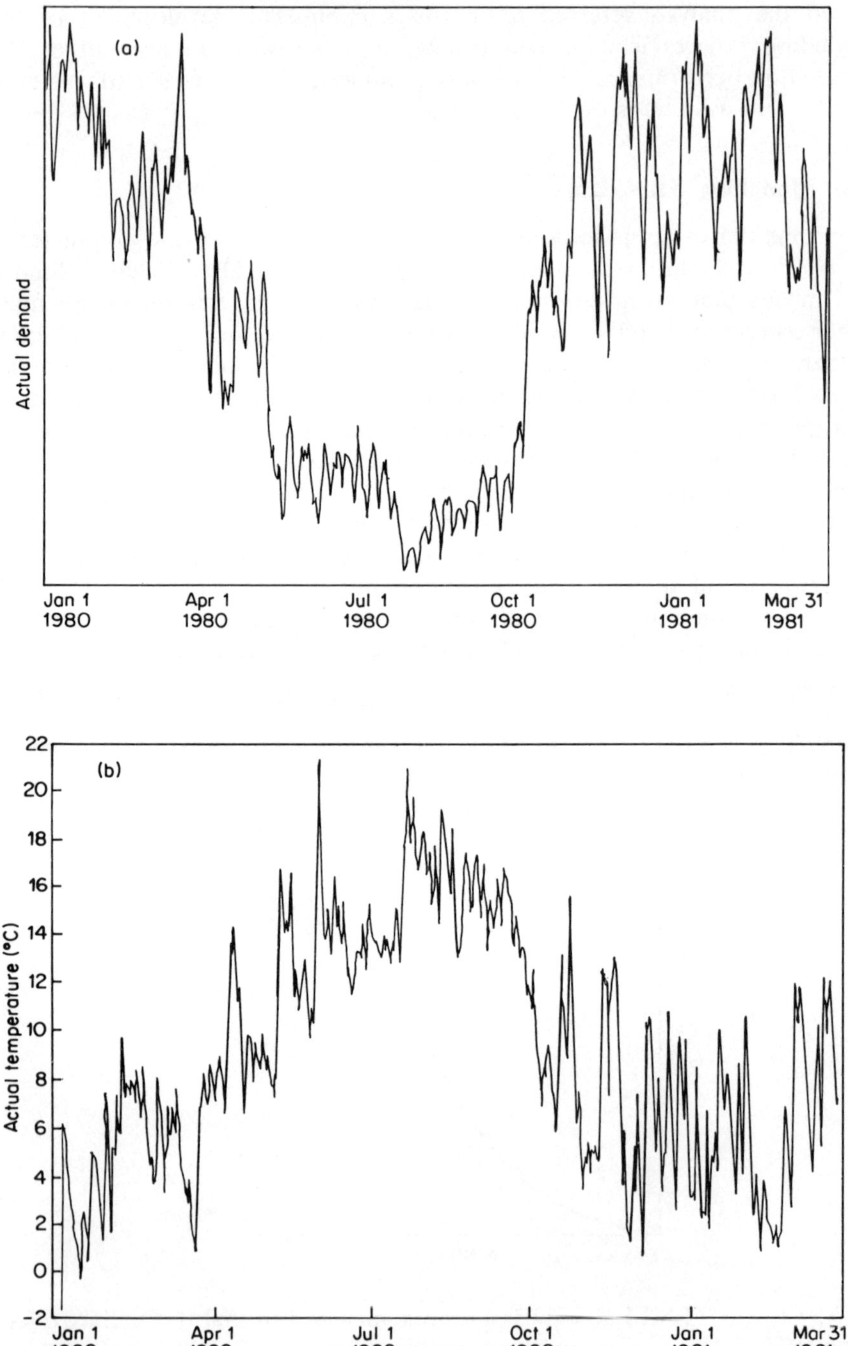
(a)
Actual demand
Jan 1 1980
Apr 1 1980
Jul 1 1980
Oct 1 1980
Jan 1 1981
Mar 31 1981
(b)
Actual temperature (°C)
22
20
18
16
14
12
10
8
6
4
2
0
-2
Jan 1 1980
Apr 1 1980
Jul 1 1980
Oct 1 1980
Jan 1 1981
Mar 31 1981

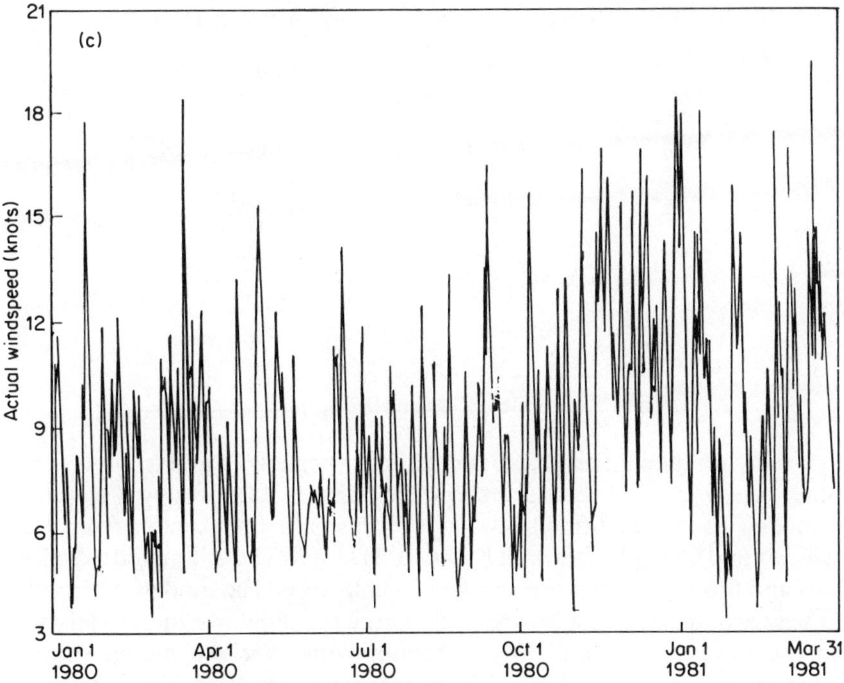

FIGURE 11.4 (a) The actual demand series; (b) the daily temperatures, Jan 1980 to Mar 1981; (c) the average daily windspeeds, Jan 1980 to Mar 1981

30 °C. This means that the chilling effect of a 20-knot wind at 0 °C is much greater than that of a 20-knot wind at 10 °C.

It is interesting to note that whichever data set is used the same two basic models are identified. These models are:

$$\nabla\nabla_7 D_t = \left(\frac{\omega_0}{1-\delta_1}\right)\nabla\nabla_7 T_t + (1-\theta_1 B - \theta_2 B^2)(1-\theta_7 B^7)\alpha_t \qquad (11.1)$$

$$\nabla\nabla_7 D_t = (\omega_0 - \omega_1)\nabla\nabla_7 T_t + (1-\theta_1 B - \theta_2 B^2)(1-\theta_7 B^7)\alpha_t \qquad (11.2)$$

where ω_0, ω_1, δ_1, θ_1, θ_2, and θ_7 are parameters to be identified, ∇ is the difference operator, and B the backward shift operator. The models look rather complicated, but the transfer function simply measures the demand/temperature response, the ∇_7 term introduces the weekly pattern relating the demand today to the demand 7 days ago, and the noise model 'mops' up any structure that is left in the data after weather effects have been removed. A more detailed description of the models is given in the Appendix.

The demand data for 1980/81 is shown in Figure 11.4(a). The data in the

TABLE 11.6 Comparison of the fit of the two best daily models

Data	Model 1 Jan–Jun	Model 2 Jan–Jun	Model 1 Sep–Jan	Model 2 Sep–Jan
ω_0	−140.2	−131.0	−144.2	−135.9
ω_1		85.2		90.0
δ_1	0.44		0.47	
θ_1	0.44	0.33	0.37	0.19
θ_2	0.27	0.27	0.29	0.33
θ_7	0.91	0.91	0.95	0.95
Residual variance	29942	31313	26422	27574
Q_{ACF}	52	70	50	51
Q_{CCF}	49	53	53	57
G_{AIN}	250	216	272	226

figure has not been corrected for holiday periods. Similar plots of the temperature and wind data are shown in Figures 11.4(b) and 11.4(c). The two models have been fitted to two sets of this data, one running from 1/1/80 to 1/6/80 and the other from 1/9/80 to 31/1/81. The estimated parameter values and fitting statistics are shown in Table 11.6. The models which fit the data best are those with a low residual variance. Similarly the models which have the lowest Q_{ACF} and Q_{CCF} account better for the model structure. Thus it can be seen that Model 1 is preferable to Model 2. These results are for data where no wind effect has been included. However, similar results are obtained when account is taken of the windspeed. To compare the effect of including windspeed, a comparision of the fitting statistics for Model 1, using temperature only, wind and temperature, and chill effect are shown in Table 11.7. This clearly indicates that a better fit is obtained when wind is included, and an even better fit is obtained when using chill. The best test of a model is its forecasting ability, which is discussed in the next section.

4.3 Forecasting daily gas demand

The usual way to test the forecasting ability of models is to model using data up to a given date, and then to analyse the forecasts after that date. This has

TABLE 11.7 Comparision of models including and excluding windspeed (Model 1)

	Jan–Jun			Sep–Jan		
	Temp	Wind	Chill	Temp	Wind	Chill
Residual variance	29942	24768	21774	26422	21213	18316
Q_{ACF}	52	45	38	50	47	42
Q_{CCF}	49	48	45	53	53	54

TABLE 11.8 Forecasting performance of temperature, wind, and chill models for June 1980 and February 1981

Model Type	Temp 1	Temp 2	Wind 1	Wind 2	Chill 1	Chill 2	
Bias %	−1.2	−1.4	−1.3	−1.8	−2.0	−2.1	
Accuracy %	4.9	4.5	5.0	4.5	5.5	4.2	June 1980
S. deviation %	5.8	5.4	5.7	5.4	6.3	5.1	
Bias %	0.09	−0.10	0.21	0.09	0.66	0.21	
Accuracy %	1.9	1.8	1.5	1.5	1.2	1.2	Feb 1981
S. Deviation %	2.3	2.3	1.8	1.8	1.4	1.4	

been done for the two models described in the previous section, using temperature, wind, and chill data. The results are shown in Table 11.8, both for June 1980, using January to May 1980 to model, and February 1981 using September 1980 to January 1981 to model.

The results for June 1980 are not very impressive, considering actual temperatures and windspeeds have been used; however, the results for February 1981 are encouraging, the chill model outperforming the wind and temperature models. It is interesting to note that there is little or no difference between the forecasting performance of Model 1 and Model 2. If the results for June 1980 are analysed more closely then it is found that most of the worst forecasts occur at weekends, especially Saturdays, the day in which the lowest demand in a week occurs. These bad forecasts are all underforecasts, indicating that at low demands the spread of the data is lower than is expected. To avoid this problem the logarithm of the demand series is taken before modelling. This has the effect of reducing the large number of underforecasts at weekends. The forecasting performance of the log models are shown in Table 11.9. These results are considerably better than those obtained previously, and show that there is very little difference between the forecasting performance of models using temperature, temperature and wind, and chill. However they do indicate that Model 1 produces consistently better forecasts than Model 2.

The use of 'log' models in the winter period, however, makes the forecasts worse. The current program therefore uses log models from mid-May until

TABLE 11.9 The forecasting performance of log models for June 1980

Data used Model	Temperature 1	Temperature 2	Temperature and wind 1	Temperature and wind 2	Chill 1	Chill 2
Bias %	0.7	−0.1	0.3	0.2	0.3	0.4
Accuracy %	2.7	2.9	2.5	2.7	2.5	2.9
S. deviation %	3.8	4.1	3.6	3.9	3.4	3.8

TABLE 11.10 Comparison of forecasts for February and March 1981 using actual and forecast temperatures

Time of forecast	1600 Day −1	0800 Day 0	Retrospectively	
Bias %	1.7	1.4	0.7	F
Accuracy %	3.2	2.0	1.2	E
S. Deviation %	3.6	2.4	1.4	B
Bias %	0.9	1.2	1.0	M
Accuracy %	4.7	2.9	2.3	A
S. deviation %	6.0	3.5	2.5	R

the end of September and non-log models for the remainder of the year. The actual model used is of Type 1, using chill as the input series.

It is interesting to compare the actual forecasts made 'on-line' using forecast temperatures with those made retrospectively using actual temperatures. Such a comparison is shown in Table 11.10, for a forecast (a) made at 1600 on the previous day, using an estimate of the demand for that day, (b) made at 0800, and (c) made retrospectively. These data are for the month of February 1981, and for March 1981.

These results indicate that at 0800 approximately 0.7% of the error is due to errors in the weather forecasts, whereas at 1600 on the previous day approximately 2.1% of the error is due to a combination of forecast temperature error and demand forecast error. (The forecast made at 1600 on the previous day depends on both the weather forecast for that day and the current day, and on the forecast demand for the previous day.) This leaves a considerable random fluctuation which is impossible to forecast. Such fluctuation may be due to other weather factors such as sunshine and rainfall not being included in the model or to the way in which individual regions use storage gas. The demand passed onto the national system is the demand plus or minus the storage used and not the true demand of the customers.

4.4 Modelling weekly demand

Temperature and windspeed affect weekly demand, as they do daily demand; however, when forecasting the demand for a few weeks ahead no forecasts of temperature or windspeed are available. For future temperatures it is possible to use seasonal normal temperatures (SNT), but no such quantity exists for windspeed. This leaves two choices when modelling: (i) to ignore temperature affects, and (ii) to include temperature. A plot of the weekly demand data for 1976–1980 is shown in Figure 11.5(a) and for the average weekly temperatures for the same period in Figure 11.5(b). Data from 1976 to 1979 has been used to produce the best 'demand only' or

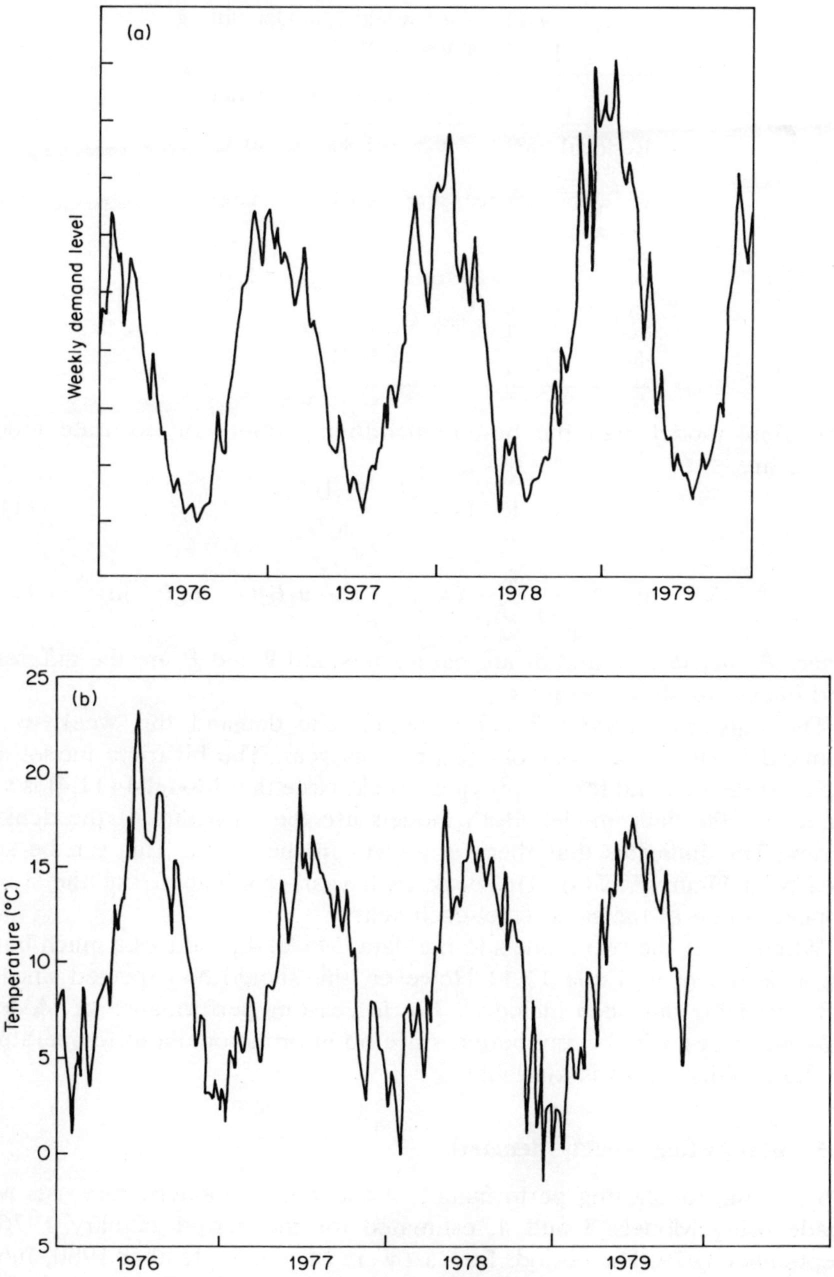

FIGURE 11.5 (a) The weekly demand, 1976–79; (b) the weekly temperatures, 1976–79

TABLE 11.11 The weekly model fitting parameters

	Model 3	Model 4
Residual sum of squares	1.84	0.32
Residual variance	0.013	0.002
Q_{ACF}	49.3	45.5
ω_0	—	0.04
δ_1	—	0.16
θ_1	—	0.72
θ_{52}	0.75	0.77
ϕ_1	0.71	—

univariate model, and the best demand/temperature or bivariate model. These are:

$$\nabla_{52} \log D_t = \frac{(1-\theta_{52}B^{52})}{(1-\phi_1 B)} a_t \tag{11.3}$$

$$\nabla\nabla_{52} \log D_t = \frac{\omega_0}{(1-\delta_1)} \nabla\nabla_{52} T_t + (1-\theta_1 B)(1-\theta_{52}B^{52})a_t \tag{11.4}$$

where θ_1, θ_{52}, ϕ_1, ω_0, and δ_1 are parameters and ∇ and B are the difference and backward shift operators.

The univariate Model 3 (11.3) relates the demand this week to the demand in the same week of the previous year. The bivariate model also includes the demand for the previous week. Note that Model 4 (11.4) is very similar to the daily model. Both models use the logarithm of the demand series. This indicates that there is growth in the series. This can be seen clearly in Figure 11.5(a). The peak each year gets higher, but the trough appears to be at the same level each year.

When fitting the two models to the data, Model 4 produces a much better fit, as is shown in Table 11.11 However, this should be expected since an extra variable has been included. The forecasting performance of Model 4 may not necessarily be any better, since no information about temperatures in the coming weeks is available.

4.5 Forecasting weekly demand

To test the forecasting performance of the weekly models, forecasts were made using Models 3 and 4, estimated for the period January 1976 to September 1979. The periods forecast were May to September 1980, July to September 1980, and September 1980. For Model 4 seasonal normal temperatures were used to produce the forecasts. The results are plotted in Figure 11.6(a) for Model 3 and Figure 11.6(b) for Model 4. If the two

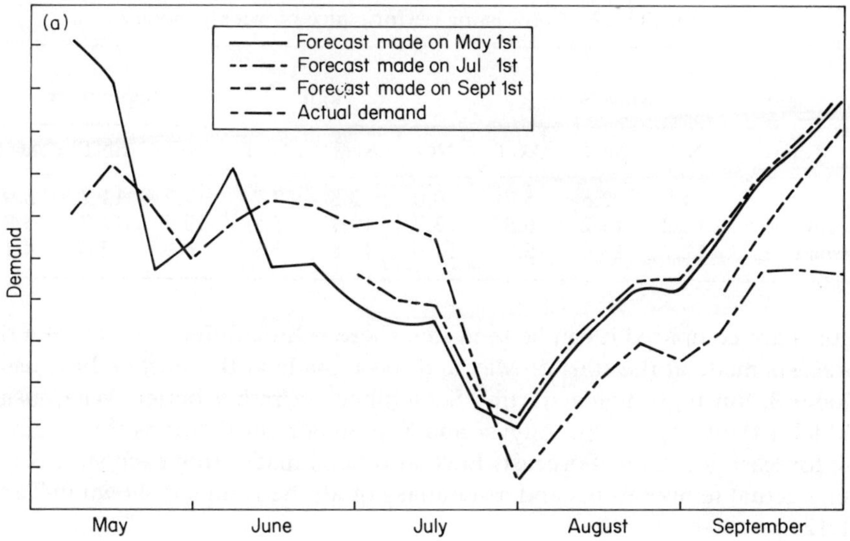

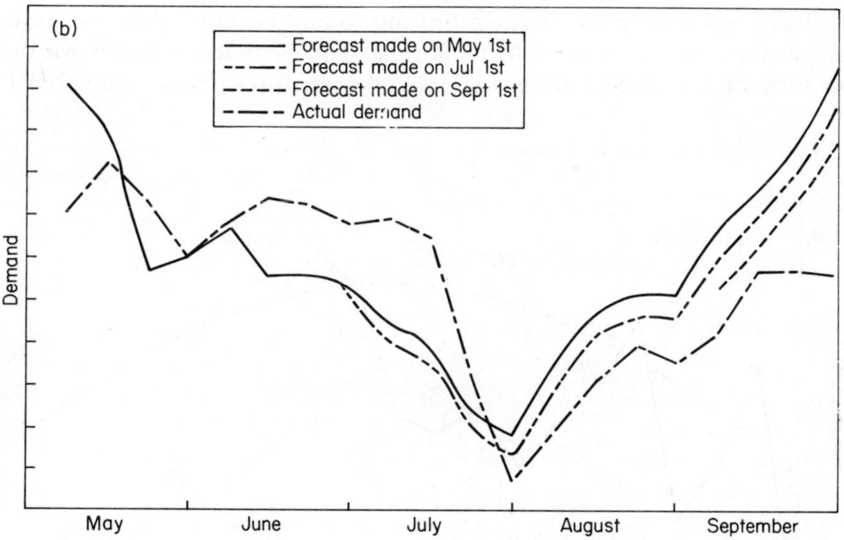

FIGURE 11.6 Forecasting performance of (a) the univariate model and (b) the bivariate model starting at different times during the year

TABLE 11.12 Forecasting performance of weekly models

Start origin	May–Sept			July–Sept			September		
Temp	NO	SNT	ACT	NO	SNT	ACT	NO	SNT	ACT
Bias	4.6	2.5	5.7	9.0	3.8	3.7	12.7	11.2	3.9
Accuracy	12.2	12.2	6.8	13.7	12.1	5.8	12.7	11.2	3.9
Spread	12.7	13.6	6.1	11.9	12.8	5.7	6.9	6.4	4.1

figures are compared it can be seen that there is little difference between the forecasts made at the start of May and those made at the start of July, using Model 3, but those made starting September are rather better. When using Model 4 the forecasts for August and September get better as the origin of the forecast gets later. Forecasts have also been made from each start origin using actual temperatures and a summary of all the results is shown in Table 11.12.

To make more use of these results it is useful to compare the actual temperatures with SNT, and such a comparison is shown in Figure 11.7 This figure shows that the temperature up to and including July was colder than normal, August was close to normal, and September rather warmer. This shows up as underforecasts in July and overforecasts in September for

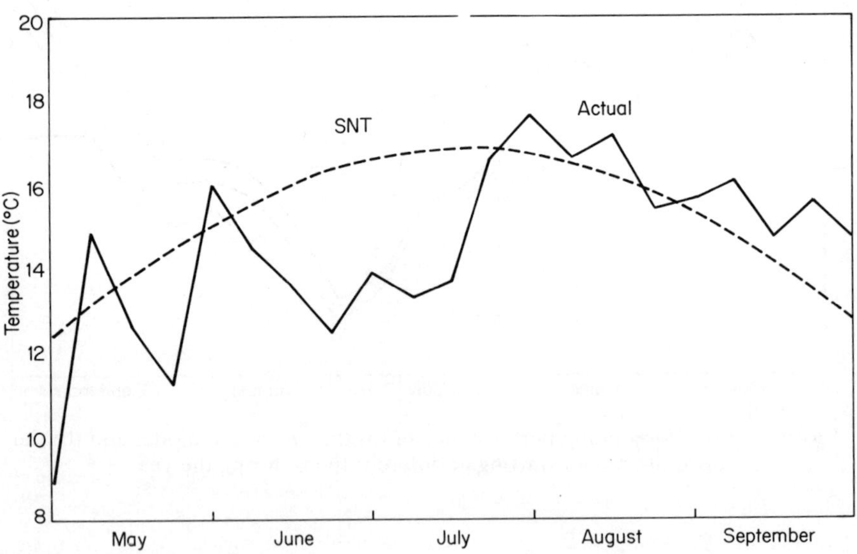

FIGURE 11.7 The actual temperature and seasonal normal temperatures for May to September 1980

both Models 3 and 4. The main conclusion is that it is slightly preferable to use a temperature/demand model, and forecasts would be much improved if a better estimate of future temperatures were to become available.

4.6 Modelling two-hourly demand

The temperature affects the two-hourly demand in a rather odd manner due to the way in which the regions take gas from the national system. Any change in demand during a day, due to changes in weather conditions, tends to be absorbed by the use of local regional storage. This means that the change is not seen on the national system until the beginning of the next day when it is seen as an abrupt change in demand level. Using individual 2-hourly temperatures is therefore almost impossible and it is better to use the average daily temperature as the input to the system. The best models identified are:

$$\nabla\nabla_{12} \log D_t = \frac{(1-\theta_1 B)(1-\theta_{12}B^{12})}{(1-\phi_{84}B^{84})} a_t \tag{11.5}$$

$$\nabla\nabla_{12} \log D_t = \left(\frac{\omega}{1-\delta_1 B}\right)\nabla\nabla_{12} T_t + \frac{(1-\theta_1 B)(1-\theta_{12}B^{12})}{(1-\phi_{84}B^{84})} a_t. \tag{11.6}$$

A plot of the two-hourly demands for 1 January 1981 to 20 February 1981 is shown in Figure 11.8. The fit of the models and parameter values are shown in Table 11.13. The data was modelled using both logarithm of the demand, and with no transform over the 7 weeks up to 31 January 1981. It can be seen from the parameter values that there is only a very weak weekly seasonality, and that the response to changes in temperature is also very small (about 1% per degree), but the sum of squares is reduced when modelling using temperature.

4.7 Forecasting two-hourly demand

When using the two-hourly demand forecasting system it is the total demand for the day rather than the individual two-hourly demands that are used. To test the forecasting ability of the models, forecasts have been made at 1600 hours and 2400 hours for the remainder of the day, and the total take for the day has been calculated. Actual temperatures have been used, and the analysis has been done for February 1981. The results are shown in Table 11.14, together with the results obtained on-line during this period.

The results indicate that log models give slightly better results than non-log models, and that using temperature as an input to the model improves the accuracy from 2.3% to 1.7% at 1600 and from 0.7% to 0.4% at 2400. The on-line model was estimated for a period several weeks

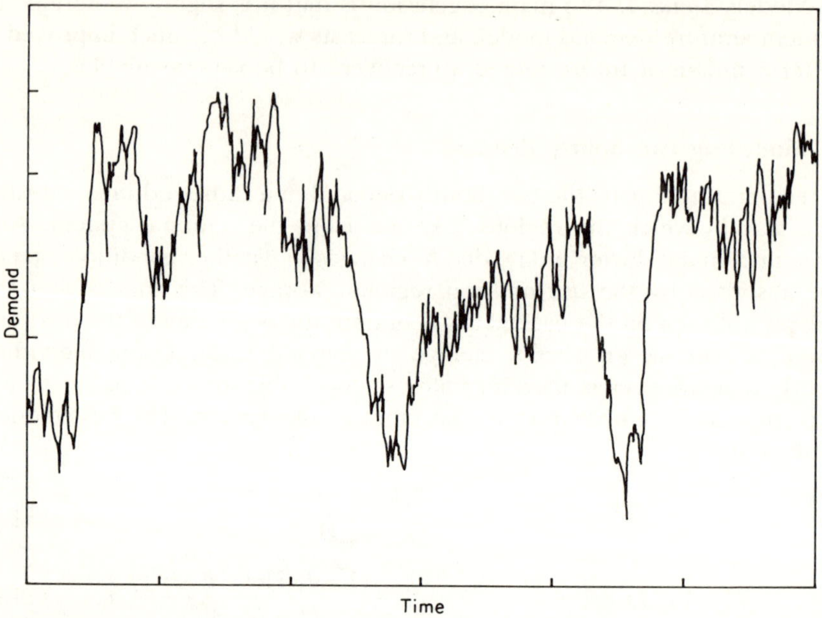

FIGURE 11.8 The two-hourly demands from 1 January 1981 to 20 February 1981

previously, and was using forecast temperatures rather than actuals. However, there may be a case for re-estimating the model parameters more frequently than is currently practised.

5 USE OF THE METGAS FORECASTING SYSTEM

The daily and within-day programs are part of an integrated system, and are run 6 times daily by the grid control engineers. At 0800 a daily forecast is

TABLE 11.13 Fitting and model parameters for within-day models

	Untransformed data		Logarithm data	
	Univariate	Temperature	Univariate	Temperature
Sum of squares	125971	118632	0.453	0.424
Q_{ACF}	50.6	42.2	47.0	46.2
ω_0	—	−4.6	—	−0.009
θ_1	0.15	0.18	0.13	0.17
θ_{12}	0.97	0.97	0.97	0.96
ϕ_{84}	0.01	0.08	0.01	0.08

TABLE 11.14 Forecasting performance of within-day models

Model	5	6	5	6	On-line	
Transform	—	—	Log	Log	—	Time
Bias %	−0.1	0.2	0.0	0.2	0.33	
Accuracy %	2.29	1.74	2.25	1.71	1.95	1600
Spread %	2.87	2.15	2.84	2.11	2.33	
Bias %	−0.0	0.1	−0.0	0.1	0.19	
Accuracy %	0.68	0.46	0.68	0.44	0.57	2400
Spread %	0.84	0.59	0.83	0.57	0.65	

run, and at 1200 and thereafter a within-day forecast is run. A within-day forecast produces a forecast demand for the current day. This demand is then used as the input to the daily forecasting program to produce a forecast for the following day. For forecasting the current day's total demand it is better to use a daily forecast up to 1200, and a within-day forecast from 1400 onwards. The weekly forecasting program is run approximately once a week, though more often towards the end of the year when scheduling of supply is more important. The program produces a forecast of the average daily demand for each week until the end of the year, together with a forecast of the total amount of gas required until the end of the year.

DISCUSSION

The forecasting system currently in use works well for most of the time; however, there are several improvements that could be made to improve the forecasts. It is well known that the demand/temperature response changes throughout the year. Until recently it has not been possible to incorporate such changes in response into the daily models, though using new software this may now be possible. The way in which the wind should be incorporated into demand forecasting models is not clear and there are other methods (e.g. including windspeed as a separate input) which are under investigation. Windspeed could also be included in the within-day models, although it is not clear what effect wind-speed has on the hourly demand. Another field in which work is being carried out is in the forecasting of the individual 2-hourly demands or daily profile, rather than forecasting just the total for the day. This profile probably depends as much on the forecast temperature a few hours ago as on the current temperature. This is due to the delay between the customer increasing demand and more gas being taken from the national system due to use of regional storage and linepack. All these possible improvements to the forecasting system are relatively minor and would not affect the basic system.

7 CONCLUSION

The automatic system for decoding weather forecasts telexed directly into the computer system has been in operation for nearly 3 years and works successfully with very little manual intervention being required. All the forecasts are stored and a large database of past forecasts and actuals has been collected. This database is easily interrogated and much use is made of the contents.

The gas demand forecasting system is used regularly by the grid control engineers for forecasting the total demand in the next few days, and for forecast demand in the coming weeks. The daily forecasting program and the weather forecast output programs are run at every forecast time, i.e. every 4 hours, and they have proved to be a useful aid to the grid control engineers in assisting them when making control decisions.

ACKNOWLEDGEMENTS

The author would like to thank British Gas Corporation for permission to publish this paper and to acknowledge the work done by M. D. Harper of British Gas London Computing Centre in setting up the telex decoding system, and to other members of London Research Station who helped to produce the results for the paper—namely C. Ringrose, A. W. Whitley, P. Davies, N. C. Garg, and S. Shah. Thanks are also due to the Grid Control engineers for the helpful comments and suggestions when designing the forecasting system.

REFERENCE

1. G. E. P. Box and G. M. Jenkins (1970). *Time series analysis: Forecasting and Control*, Holden-Day, San Francisco.

APPENDIX IDENTIFICATION OF THE DAILY DEMAND MODEL

In this section, the detailed steps involved in modelling the data are described; in a sense it is a historical record of the various improvements that have been made to the original demand/temperature model. Data for the period 1 January 1980 to 1 June 1980 has been used as an example.

Modifications to the basic data

Before modelling can begin certain anomalies such as holiday effects must be removed from the data. It is known that the demand for gas decreases around all Bank Holidays, especially around Christmas and New Year. For most holidays the decrease is a similar percentage from year to year, and is

TABLE A11.1 Holiday correction factors used to produce corrected demand series

	Wed	Thur	Fri	Sat	Sun	Mon	Tue	Wed
New Year (1980)	—	1.20	1.20	1.15	1.16	1.20	1.20	1.05
Easter	1.04	1.07	1.13	1.07	1.12	1.18	1.05	1.02
Mayday	1.00	1.00	1.00	1.03	1.00	1.08	1.00	1.00
Spring	1.00	1.03	1.05	1.00	1.04	1.19	1.07	1.03
Summer	1.00	1.03	1.05	1.03	1.08	1.23	1.07	1.03
Xmas (1979)	1.00	1.00	1.08	1.11	1.20	1.30	1.35	1.35

easily estimated. For other holdays such as Christmas, which occurs on a different day of the week each year, an estimate of the decrease is made using past experience. The current values used for the holiday effects are shown in Table A11.1, and these have been used to correct the demand series.

The solid line in Figure A11.1 shows the corrected demand series, and the dotted line shows the basic demand series. Figure A11.2 shows the temperature series. Ideally the holiday effects would be estimated within the modelling program by using Box–Jenkins intervention analysis. However, at the time of writing only a single-input transfer function model is allowed in the modelling program. In the future it is hoped that such an intervention model can be fitted.

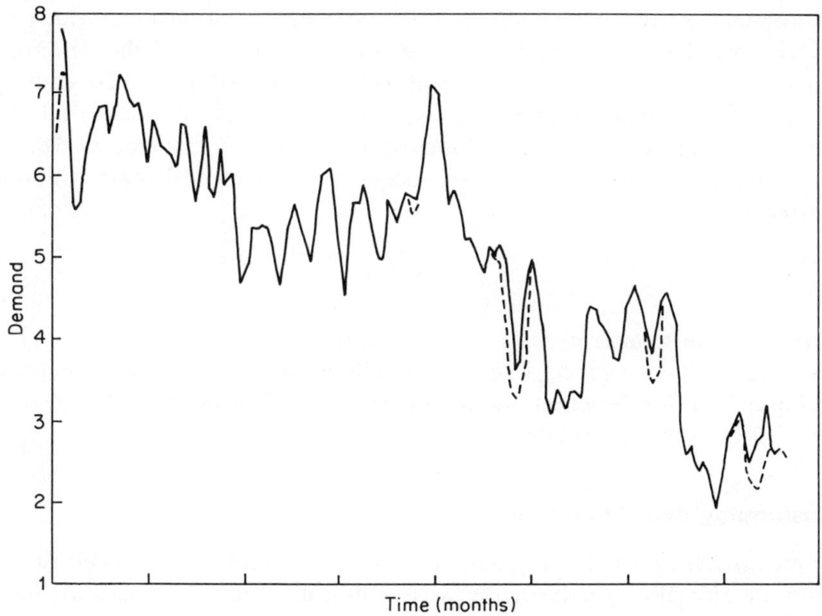

FIGURE A11.1 The demand series

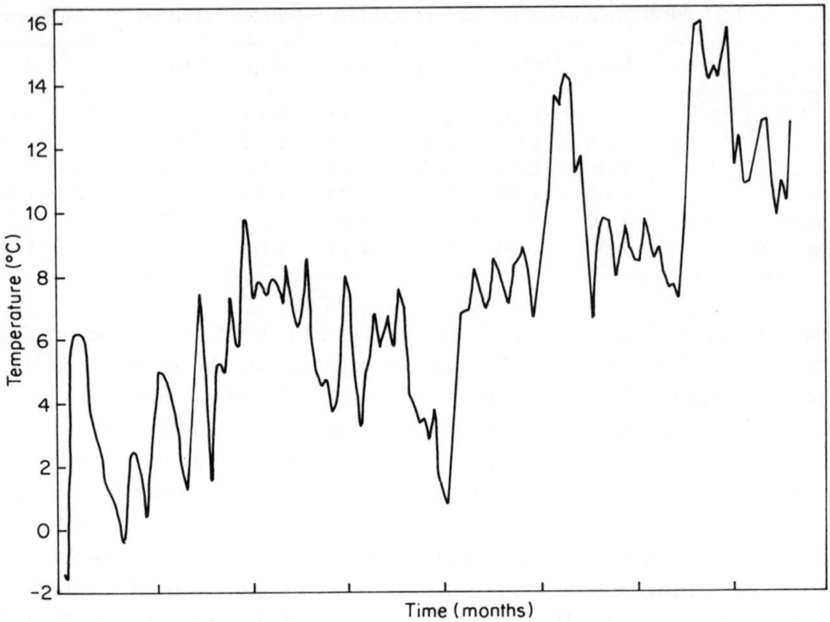

FIGURE A11.2 The temperature series

Another feature of the demand/temperature relationship, which has proved a problem in the past, is the decreasing sensitivity of the demand to changes in temperatures, for temperatures in excess of 14 °C. To allow for this effect the transform shown in equation A11.1 is applied to all daily temperatures above 14 °C. This function has a maximum value of 18. It is totally empirical, and has been used successfully for several years within the industry.

$$T_{\text{transformed}} = 13 + \frac{(1 - 0.8^{(T-13)})}{0.2} \qquad \text{(A11.1)}$$

where T is the temperature before transformation.

With these two corrected series modelling can begin, and since only a one-input transfer function modelling program is available, the effect of wind will be ignored initially.

Transforming and differencing

The mean/range plots for the data are shown in Figure A11.3. The random scatter on the plots for both the series indicate that no transforms (other than those already applied) are required.

In order to determine whether differencing is required it is necessary to

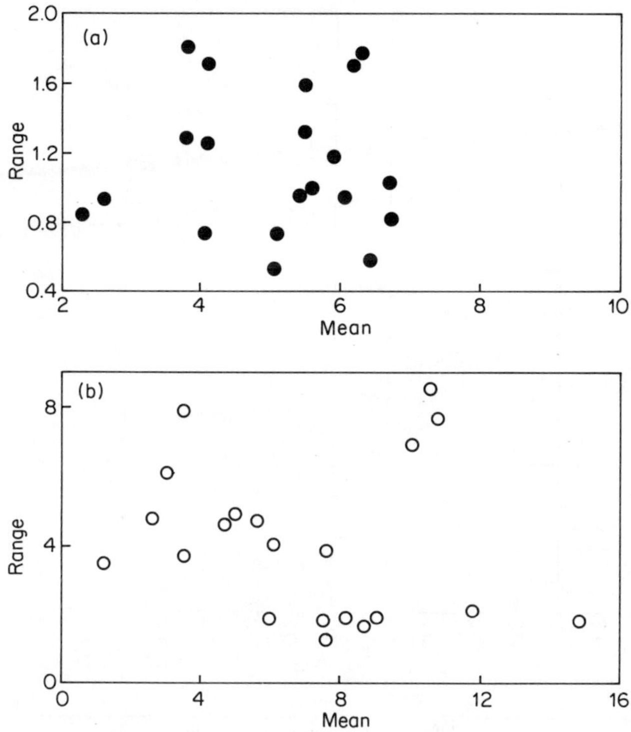

FIGURE A11.3 Mean/range plots: (a) demand; (b) temperature

look at the autocorrelation function (ACF) of the demand data. The ACF of the basic data is shown in Figure 11.4(a), the ACF after differencing once in Figure A11.4(b), and after differencing both seasonally once and nonseasonally once in Figure A11.4(d). From these plots it can be seen that a stationary series is obtained if the series is differenced once seasonally of period 7 and once nonseasonally. This differencing operation is applied both to the demand and the temperature series before prewhitening the temperature series.

Prewhitening and identification of the transfer function

Since identifying the prewhitening model is merely a device used to obtain the cross-correlation function in order to identify the transfer function, no details will be given of this identification. The reasons for prewhitening are described fully by Box and Jenkins[1] and the identification of the prewhitening model is very similar to the identification of the noise model. From the ACF and partial autocorrelation functions (PACF) of the differenced tem-

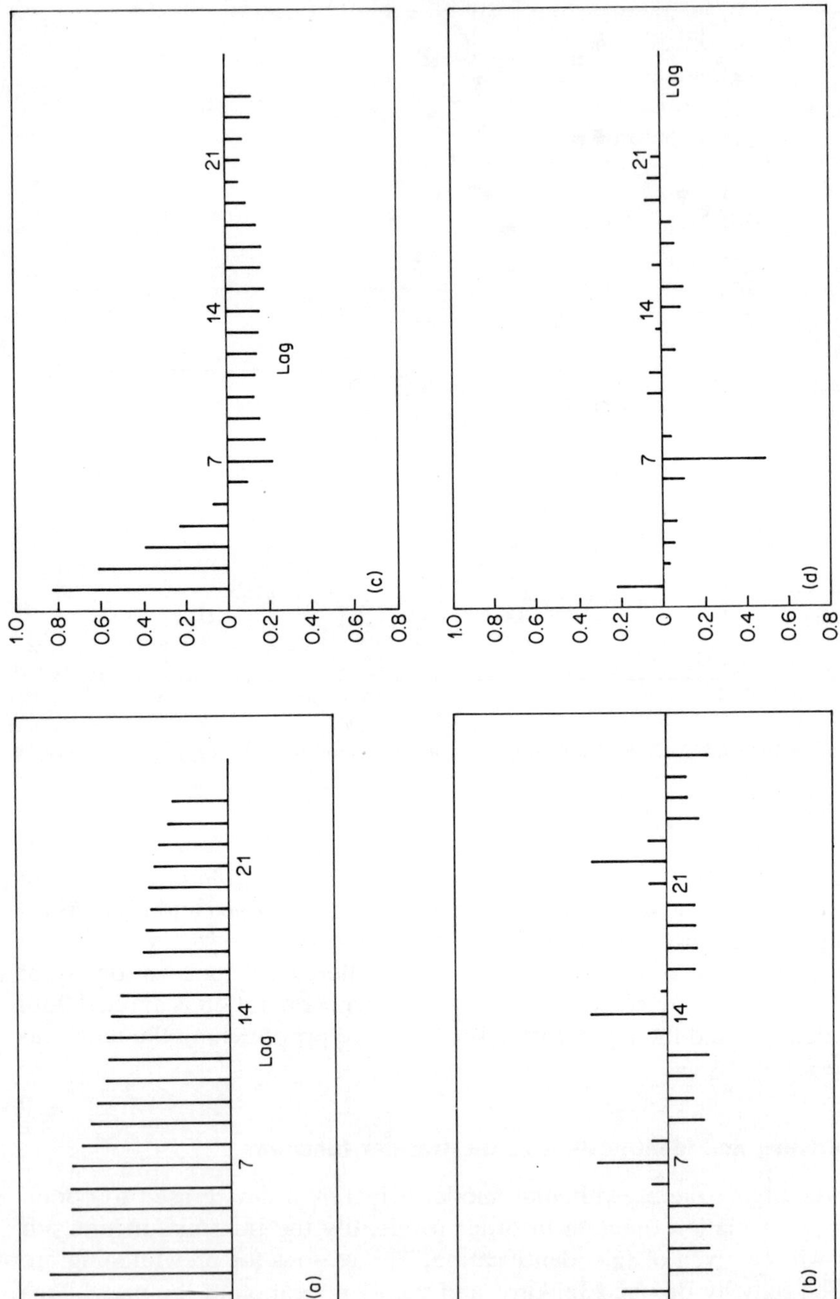

FIGURE A11.4 (a) ACF of original data; (b) ACF of demand data (after single nonseasonal differencing); (c) ACF of demand data (after seasonal differencing); (d) ACF of demand data (after both seasonal and nonseasonal differencing)

perature series, an adequate prewhitening model is found to be:

$$\nabla = \nabla_7 T = (1 - \theta_1 B - \theta_2 B^2)(1 - \theta_7 B^7)a_t$$

where T is the temperature, a_t is white noise, and θ_1, θ_2, θ_7 are parameters. The values of the parameters are:

$$\theta_1 = -0.09 \pm 0.08$$
$$\theta_2 = 0.27 \pm 0.08$$
$$\theta_7 = 0.93 \pm 0.02.$$

The θ_1 parameter is really insignificant, but is kept in order to make the model appear physically more meaningful. This prewhitening model is applied to both the demand and the temperature series before the cross-correlation function (CCF) of the two series is calculated.

The CCF is shown in Figure A11.5, in which there are only two significant residuals, those at lag 0 and lag 1. Two possible transfer function models are suggested by this structure: ARMA (1, 0) and ARMA (0, 1), as shown in equation (A11.2) and (A11.3) respectively:

$$\nabla\nabla_7 D = \frac{\omega_0}{(1 - \delta_1 B)} \nabla\nabla_7 T + N_t \qquad \text{(A11.2)}$$

$$\nabla\nabla_7 D = (\omega_0 - \omega_1 B)\nabla\nabla_7 T + N_t \qquad \text{(A11.3)}$$

where D is the demand, T the temperature, N the transfer function residuals, and ω_0, ω_1, δ_1 are parameters. Since both these models are

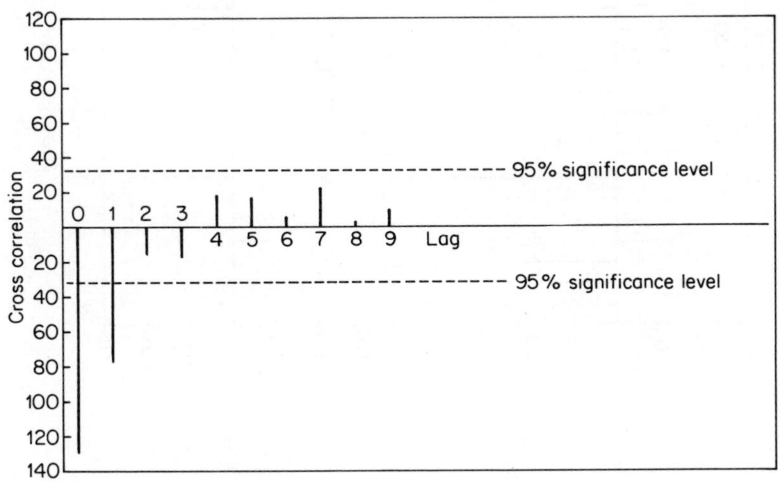

FIGURE A11.5 Cross-correlation function between prewhitened temperature and demand series

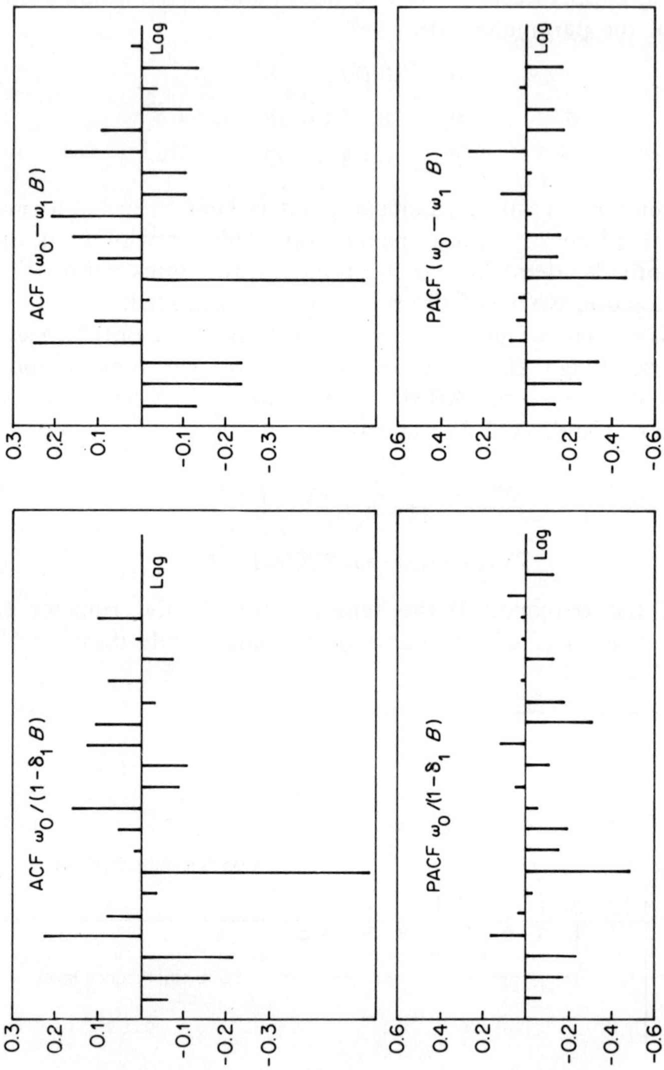

FIGURE A11.6 ACFs and PACFs of transfer function residuals

physically sensible, both are fitted to the data and the results compared. The transfer function residuals usually contain some correlation structure and it is therefore necessary to remove this structure by identifying a noise model.

Identification of the noise model

The noise model for each transfer function must be identified separately; therefore the ACF and PACF of each of the transfer-function residuals are shown in Figure A11.6. In fact the two ACFs are very similar as are the two PACFs. From the ACF and PACF patterns it is not obvious what model structure is most suitable. For similar data, a model of the form

$$N_t = (1 - \theta_1 B - \theta_2 B^2)(1 - \theta_7 B^7)a_t \qquad \text{(A11.4)}$$

shown in equation (A11.4) has been found to be adequate and in this particular case the same model form has been used.

The parameters after estimation are shown in Table A11.2 together with the variance of the residuals after fitting, and the gain of the models. The gain is calculated by replacing the operator B in the transfer function with the value 1.

There are no large correlations between the parameter estimates, and the noise model parameters are very similar for the two different transfer function models, as is the main transfer function parameter ω_0. The ARMA (1, 0) model gives a smaller variance than the ARMA (0, 1) model, which indicates that it is a slightly preferable model. The ARMA (1, 0) model also has a larger gain, which indicates that it is more sensitive to temperature changes.

Model verification

In order to check the models' suitability, it is useful to look at the ACF and PACF of the residuals after estimation. Plots of these are shown in Figure

TABLE A11.2 Final model parameter values after estimation

Parameter	Model 2	Model 3
ω_0	-140.2 ± 7.5	-131.0 ± 9.3
ω_1	—	84.9 ± 12.7
δ_1	0.44 ± 0.04	—
θ_1	0.04 ± 0.08	0.33 ± 0.08
θ_2	0.27 ± 0.08	0.27 ± 0.08
θ_7	0.91 ± 0.02	0.91 ± 0.02
Variance	29942	31313
Gain	250.4	215.9

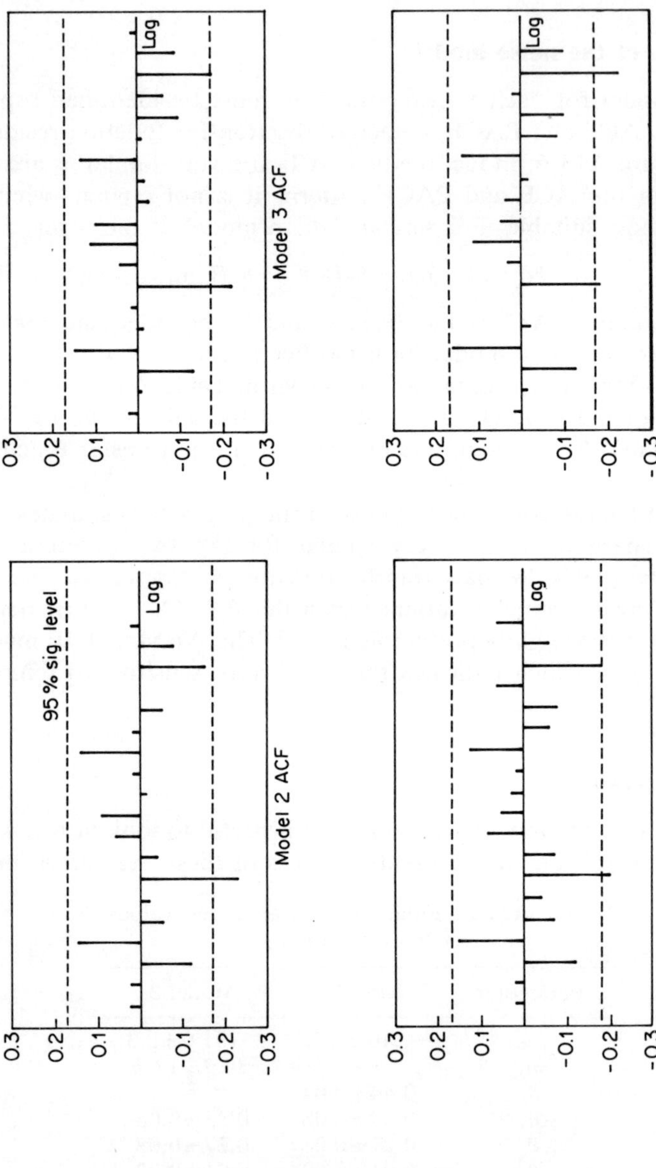

FIGURE A11.7 The ACF and PACF of the noise model residuals (after estimation for both unidentified transfer-function models)

A11.7. The Q-statistics for the ACF of Models 2 and 3 are 51.9 and 70.1 respectively both for 44 degrees of freedom, again indicating that the ARMA (1, 0) transfer function is the better model. Another useful indicator of model suitability is the cross-correlation function between the prewhitened input series and the noise residuals.[1] These CCFs are plotted in Figure A11.8, again indicating that the ARMA (1, 0) model gives the better fit. Figure A11.9 shows the ARMA (1, 0) model residuals after estimation. The residuals from the ARMA (1, 0) are very similar, having a slightly larger standard error. The figure shows that there is only one part of the series where the residuals are large for several days. This is around day 88, which turns out to be Good Friday. The negative residuals indicate that the holiday correction is too small for this particular Easter. The large residual at day 141 is also due to an inadequate holiday factor, in this case Spring

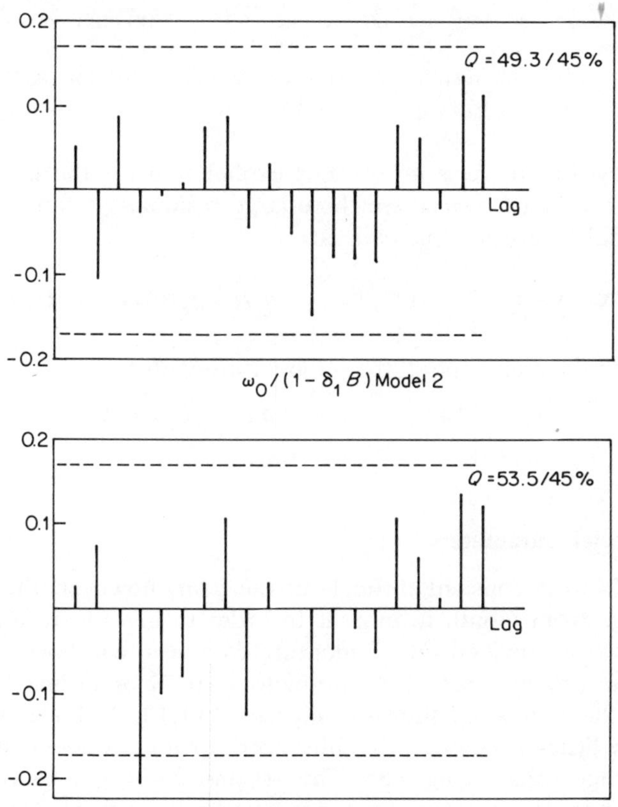

FIGURE A11.8 Cross-correlation function (between the prewhitened input and noise residuals)

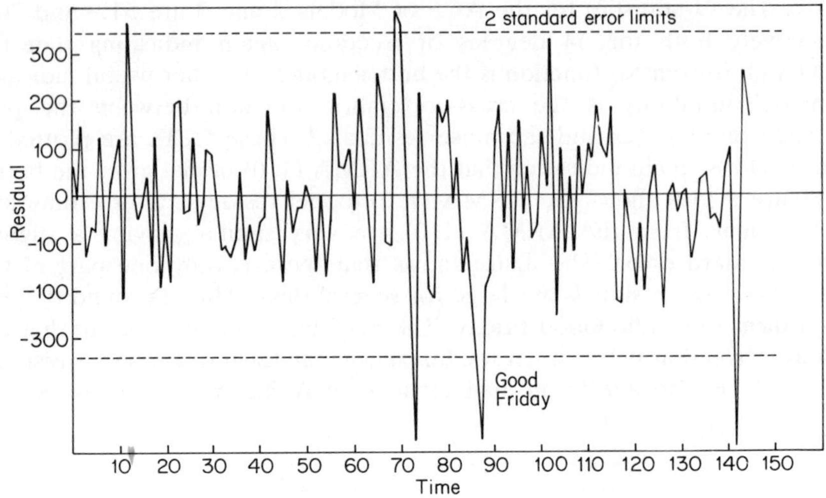

FIGURE A11.9 The residual series after estimation for the ARMA (1, 0) model

Bank Holiday, but there is no obvious explanation for the large residuals around day 71. Apart from these three large residuals, it can be concluded that the model to use for this data is:

$$\nabla\nabla_7 D = \frac{\omega_0}{(1-\delta_1 B)} \nabla\nabla_7 T + (1-\theta_1 B - \theta_2 B^2)(1-\theta_7 B^7)a_t$$

where D is the demand and T the temperature with

$$\omega_0 = -140.2 \qquad \delta_1 = 0.44 \qquad \theta_1 = 0.40$$
$$\theta_2 = 0.27 \qquad \theta_7 = 0.91.$$

Unstable model parameters

The model form is constant throughout the year, however, the parameter values change from month to month. In order to investigate this variation the data has been divided into 2-monthly data sets, each overlapping one month of the previous set. The parameters are then estimated for each period, and the results are shown in Figures A11.10, A11.11, and A11.12. There is very little variation in the 7th-order moving average noise parameter, the average value being 0.85. The 1st and 2nd-order moving average parameters show much more variability, though there is no distinct pattern visible except that the parameters are almost insignificant during the late summer and early autumn. Unlike the noise parameters, the transfer function parameters, especially the constant term, do show a cyclic pattern. The

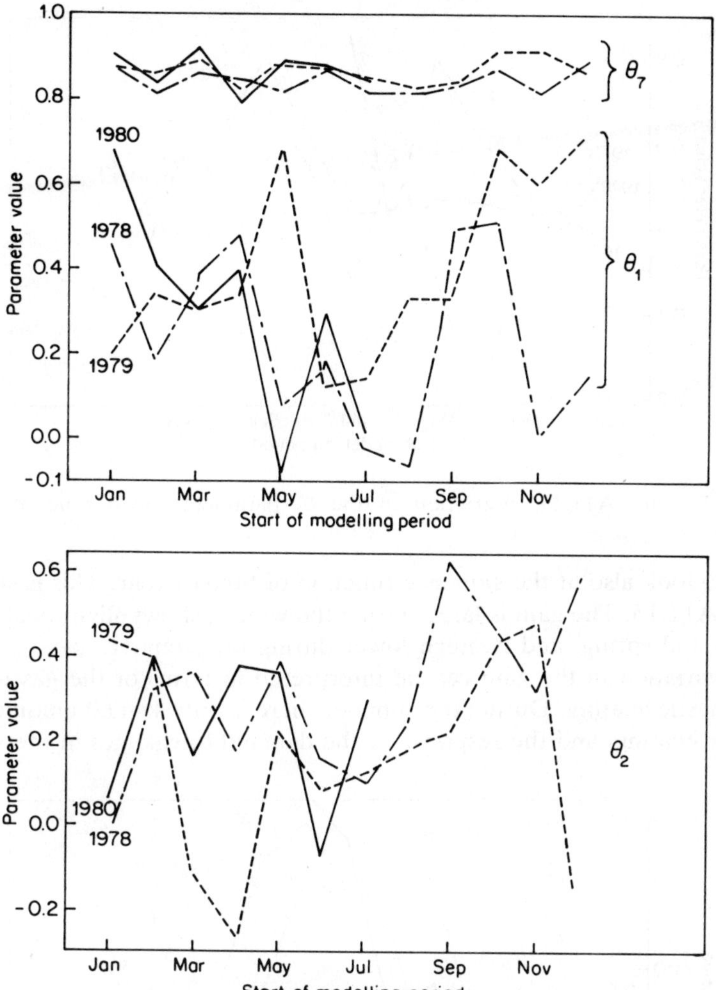

FIGURE A11.10 Variation of the moving average noise parameters (with time of year)

autoregressive parameter is reasonably constant, except for one small and one large value occurring during spring 1978 and 1979 respectively, and for several large values occurring during the autumn of 1978 and 1979. The constant term shows a very distinct pattern having large negative values during the winter and small negative values during the summer. There is also some evidence for the largest negative values occurring during the spring and least negative values during the early autumn. Since it is the transfer function parameters which show the more distinct pattern it is

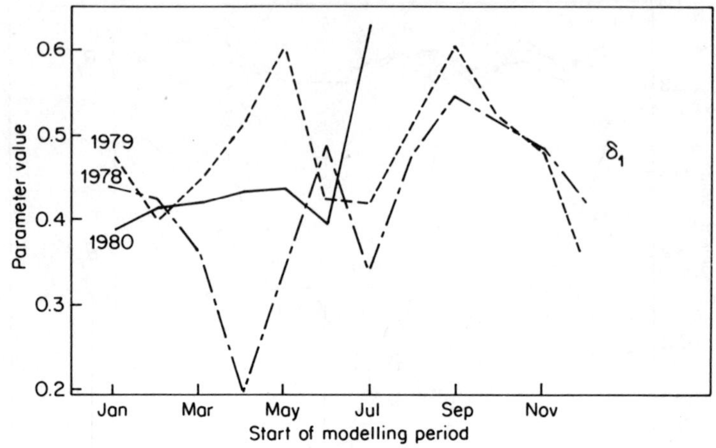

FIGURE A11.11 Variation of the δ_1 parameter (with time of year)

useful to look also at the gain as a function of time of year. This is shown in Figure A11.13. The gain is large during the winter, shows slight peaks in the autumn and spring, and is much lower during the summer.

The variation in the gain can be interpreted in terms of the gas required for domestic heating. During the summer there is only a small amount of gas used for heating, and the response of the demand to changes in the weather

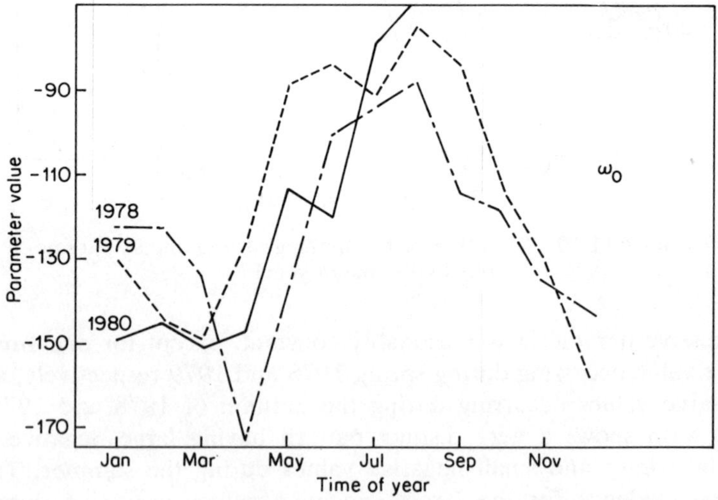

FIGURE A11.12 Variation of the ω_0 parameter (with time of year)

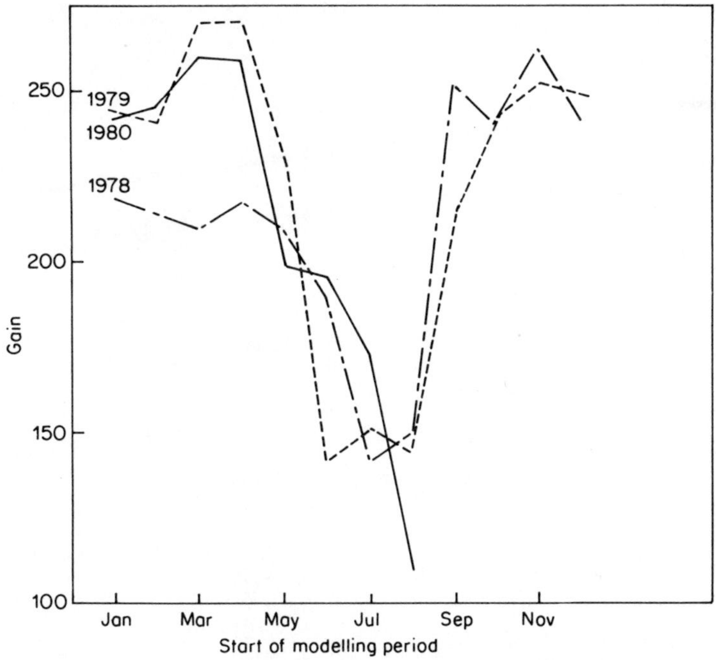

FIGURE A11.13 Variation of the gain (with the time of year)

is small. During the spring and autumn gas users are sensitive to weather changes, continually switching both fires and central heating systems on and off. This causes a large change in the demand in response to temperature changes. In the winter most central heating systems are automatically controlled, and gas fires are often switched on simply because it is winter, and the demand sensitivity to temperature decreases slightly from the autumn and spring peaks.

Currently no account is taken of this variability in the parameter estimates. It is found that by using a data series of 5 months to estimate the parameter values they become more stable. This leads to fewer large forecasting errors, and the models are more robust. Each 5-monthly model is used to forecast for the next 2 months, before the parameters are re-estimated.

Note:

This chapter is an extended version of the same author's paper 'The METGAS System' *Gas Engineering and Management*, 1982, copyright Institution of Gas Engineers, published here with permission.

Comparative Models for Electrical Load Forecasting
Edited by D. W. Bunn and E. D. Farmer

CHAPTER 12

Adaptive Forecasting of Daily Water Demand

M. J. H. Sterling and A. Bargiela

1 INTRODUCTION

The planning and operation of water supply systems has become a complex task which inevitably relies on a forecast of the likely consumer demand. The availability of a long-term forecast over several years enables a planned enhancement of the existing network with sufficient lead time to facilitate major construction work. In the short term the network resources can be regarded as fixed but the method of operation can be varied to minimize costs while satisfying all network constraints. The solution of the resultant optimization problem normally necessitates an on-line computer control system which provides for a comprehensive network monitoring scheme together with appropriate control actuations. Data is consequently available on the actual water consumption at various take-off points on every measurement scan which can then be assembled to provide an aggregated historical profile of the demand over the whole network or within selected areas. An examination of this data will normally reveal a cyclic pattern with similarities between successive days, weeks, and months which can form the basis of an on-line short-term forecasting technique. Such techniques have been investigated for prediction of consumer demand in other public utilities including the electricity supply industry,[1–6] but rather less attention has been paid to demand prediction for the water industry.

Essentially the available data on which to base a short-term forecast are a log of past consumptions with corresponding network configurations together with a meteorological record and weather forecast. Methods utilizing some of this information have been developed,[7,8] but normally require comprehensive on-line data collection of parameters other than those provided by conventional monitoring and control installations. Consequently methods which rely primarily on a log of past consumption only are likely to be more amenable for on-line demand prediction.[2]

This chapter considers the methodology of one such class of demand modelling and forecasting techniques, namely that of time series analysis based on a Box–Jenkins[9] form of autoregressive moving average demand model. The chapter considers the development of a suitable model from a nonstationary regular time series representation of consumption data and shows how such models may be evaluated, improved, and used for prediction. A new matrix formulation of the problem is developed and a novel application of the Fletcher–Powell Variable Metric Algorithm is shown to result in a unity rank updating formula which gives a very efficient, self-adaptive algorithm for on-line demand prediction.

2 DATA TRANSFORMATION

The water demand over a particular period can be characterized by a series of discrete values $[x_t, t=1, 2, \ldots, n]$. The objective of demand prediction is consequently to establish the values of the series x_t for $t=n+1, n+2, \ldots$. The series x_t will in general be nonstationary since its mean value will not be constant over any period of time. Consequently the stochastic model used to represent x_t must either allow for nonstationarity or the data must be transformed to generate a new time series which is stationary.

The transformation of the demand data x_t should not only produce a stationary series but should also be devised such that it produces either a constant error variance or a normal error distribution. Transformed data differencing may also be required to ensure stationarity, and the period can be estimated from the sample autocorrelation function (SAF). Strong correlation between samples at some interval apart would indicate that differencing should be applied with this period. The data transformation may be expressed as

$$z_t = f(x_t) \tag{12.1}$$

where f is the particular transformation for the demand data.

The differencing operation is then represented by

$$w_t = (1-B)^d (1-B^s)^D z_t$$

where d, D, and s are integers and B is the backward difference operator such that

$$(1-B)z_t = z_t - z_{t-1}$$

and s is the seasonal periodicity of the time series z_t. Now defining

$$\nabla^d = (1-B)^d \quad \text{and} \quad \nabla s^D = (1-B^s)^D$$

then the stationary time series w_t is given by

$$w_t = \nabla^d \, \nabla s^D z_t. \tag{12.2}$$

It will rarely be necessary to use values d and D greater than three to achieve stationarity.

3 SELECTION OF THE MODEL STRUCTURE

The prediction problem has now been reduced to the determination of a class of models which will adequately represent stationary rather than nonstationary time series. One such class of models is composed of autoregressive (AR) and moving average (MA) components given by

$$\phi(B)\Phi(B^s)w_t = \theta(B)\Theta(B^s)a_t \qquad (12.3)$$

where ϕ, Φ, θ, and Θ are polynomials in B such that

$$\phi(B) = 1 - \phi_1 B - \phi_2 B^2 - \ldots - \phi_p B^p \qquad (12.4)$$

$$\Phi(B^s) = 1 - \Phi_1 B^s - \Phi_2 B^{2s} \ldots - \Phi_P B^{Ps} \qquad (12.5)$$

are the autoregressive components and

$$\theta(B) = 1 - \theta_1 B - \theta_2 B^2 - \ldots - \theta_q B^q \qquad (12.6)$$

$$\Theta(B^s) = 1 - \Theta_1 B^s - \Theta_2 B^{2s} - \ldots - \Theta_Q B^{Qs} \qquad (12.7)$$

are the moving average components. If p, P_q, and Q are correctly determined then a_t will be a white noise sequence distributed as $N(0, \sigma_a^2)$ where σ_a^2 is the variance.

To ensure that equation (12.3) represents a stationary time series the coefficients of (12.4) and (12.5) must all lie outside the unit circle, hence

$$-1 < \phi_i < 1 \qquad i = 1, 2, \ldots, p$$

$$-1 < \Phi_j < 1 \qquad j = 1, 2, \ldots, P.$$

Furthermore, to ensure invertibility the coefficients of equations (12.6) and (12.7) must also all lie outside the unit circle, hence

$$-1 < \theta_k < 1 \qquad k = 1, 2, \ldots, q$$

$$-1 < \Theta_i < 1 \qquad l = 1, 2, \ldots, Q.$$

The autoregressive moving average model class defined by equation (12.3) is extensive and it is consequently necessary to establish which parameters are significant for the data to be modelled. The sample autocorrelation function can be used as an indication of the subset of models likely to be relevant, and is defined for a time series w_t as

$$r_k(w_t) = c_k(w_t)/c_0(w_t) \qquad (12.8)$$

where

$$c_k(w_t) = \frac{1}{n}\sum_{t=1}^{n-k}(w_t - \bar{w})(w_{t+k} - \bar{w}) \qquad (12.9)$$

and

$$\bar{w} = \frac{1}{n} \sum_{t=1}^{n} w_t. \tag{12.10}$$

The value of r_k can thus be calculated from the data w_t for $k = 0, 1, \ldots, n-1$ and will indicate the relationship between samples k periods apart. Values of $r_k(w_t)$ which are close to zero imply that samples k periods apart are not significantly related, whereas a SAF approaching unity indicates a strong interdependence of samples at that frequency. The threshold for parameter significance is normnally defined in terms of the standard error (SE) given by

$$\mathrm{SE}(r_k) \simeq \frac{1}{n}\left(1 + 2\sum_{j=1}^{q} r_j^2\right) \qquad k > q \tag{12.11}$$

where q is the value of k beyond which the SAF is small, such that if $r_k(w) > 2\mathrm{SE}(r_k)$ for multiples of k, then a differencing parameter with period k should be included in the model. Further aids in the parameter identification process include the partial autocorrelation function and fast Fourier transforms.

4 MODEL PARAMETER FITTING

Having established the form of the autoregressive moving average model required for the data, the coefficients associated with each backshift operator must be determined. The objective function normally associated with this fitting procedure is a minimization of the sum of squared errors for each sample point. Thus the coefficients ϕ, Φ, θ, and Θ which minimize

$$S = \sum_t a_t^2(\phi, \Phi, \theta, \Theta) \tag{12.12}$$

complete the time series model.

In order to produce a computationally efficient algorithm a special matrix formulation of the problem has been developed. Writing equation (12.3) at each time step $t = 1, \ldots, N$ the model can be represented as

$$F D \mathbf{w} = R T \mathbf{a} \tag{12.13}$$

where

$$D = \begin{bmatrix} 1 & & & & & \\ \phi_1 & 1 & & & & \\ \phi_2 & \phi_1 & 1 & & & \\ & \cdot & \cdot & \cdot & & \\ & & \cdot & \cdot & \cdot & \\ & & \phi_2 & & \phi_1 & 1 \end{bmatrix} \qquad F = \begin{bmatrix} 1 & & & & & \\ & \cdot & & & & \\ & & \cdot & & & \\ \Phi_1 & \ldots & 1 & & & \\ & \cdot & & \cdot & & \\ & & \cdot & & \cdot & \\ & & & \Phi_1 & \ldots & 1 \end{bmatrix}$$

(In F, the first s rows are marked by a brace labelled s.)

$$
T=\begin{bmatrix}
1 &&&&& \\
\theta_1 & 1 &&&& \\
\theta_2 & \theta_1 & 1 &&& \\
\cdot & \cdot & \cdot &&& \\
& \cdot & \cdot & \cdot && \\
&& \cdot & \cdot & \cdot & \\
&&& \theta_2 & \theta_1 & 1
\end{bmatrix}
\qquad
R=\begin{array}{l}
s\left\{\begin{array}{c} \\ \\ \\ \\ \end{array}\right. \\ \\ \\ \\ \\
\end{array}
\begin{bmatrix}
1 &&&&& \\
& \cdot &&&& \\
&& \cdot &&& \\
&&& \cdot && \\
\Theta_1 & \cdots & 1 &&& \\
& \cdot && \cdot && \\
&& \cdot && \cdot & \\
&&& \cdot && \cdot \\
&&& \Theta_1 & \cdots & 1
\end{bmatrix}
$$

$$\mathbf{w}^{\mathrm{T}}=[w_1, w_2, \ldots, w_N]$$

$$\mathbf{a}^{\mathrm{T}}=[a_1, a_2, \ldots, a_N].$$

The function S to be minimized is now

$$S=\mathbf{w}^{\mathrm{T}}F^{\mathrm{T}}D^{\mathrm{T}}(R^{-1})^{\mathrm{T}}(T^{-1})^{\mathrm{T}}T^{-1}R^{-1}DF\mathbf{w}. \tag{12.14}$$

Using matrix E_i defined as

$$E_i=\left[\begin{array}{c:c} \multicolumn{2}{c}{0_{iN}} \\ \hdashline I_{N-1} & 0_{N-i,i} \end{array}\right] \qquad 1\leqslant i\leqslant N-1$$

the necessary conditions for this optimization problem can be written in a concise form:

$$-2\mathbf{a}^{\mathrm{T}}E_iR^{-1}T^{-1}F\mathbf{w}=0 \qquad 1\leqslant i\leqslant p \tag{12.15}$$

$$-2\mathbf{a}^{\mathrm{T}}E_{si}R^{-1}T^{-1}D\mathbf{w}=0 \qquad 1\leqslant i\leqslant P \tag{12.16}$$

$$-2\mathbf{a}^{\mathrm{T}}E_iT^{-1}\mathbf{a}=0 \qquad 1\leqslant i\leqslant q \tag{12.17}$$

$$-2\mathbf{a}^{\mathrm{T}}E_{si}R^{-1}\mathbf{a}=0 \qquad 1\leqslant i\leqslant Q. \tag{12.18}$$

These equations can be solved with respect to model parameters in several different ways, for example by using any appropriate hill-climbing method. However, in order to take full advantage of the problem structure the Newton-Raphson iterative procedure, modified to overcome numerical problems, is shown below to result in a highly efficient solution procedure.

In a standard Newton–Raphson process the corrections of the parameter estimates are calculated according to the formula

$$\Delta\Psi=H^{-1}h(\Psi) \tag{12.19}$$

where

$\Psi^{\mathrm{T}}=[\phi_1, \ldots, \phi_p, \Phi_1, \ldots, \Phi_P, \theta_1, \ldots, \theta_q, \Theta_1, \ldots, \Theta_Q]$

H^{-1} is the inverse of the Hessian matrix

$h(\Psi)$ is the gradient of the function S

the biggest computational effort being associated with the calculations of the inverse of the Hessian matrix. Additionally, if the initial estimates of parameters Ψ are far from their optimal values, equation (12.19) may prove to be numerically unstable because of an approximate singularity of the computed Hessian matrix.

These difficulties can be remedied by calculation of the corrections of the parameter estimates using an approximation of the inverse of the Hessian according to the Fletcher–Powell method. Taking Ψ_i as the current parameter estimates, the algorithm for updating model parameters may be expressed as:

(i) $\Psi_{i+1} = \Psi_i + \lambda_i v_i$ where $v_i = Q_i g_i$, and λ_i is chosen to minimize $S(\Psi_i + \lambda_i v_i)$

(ii) $u_i = \lambda_i v_i$ and $y_i = g_{i+1} - g_i$ where g_{i+1} is the new gradient vector (evaluated at Ψ_{i+1})

(iii) $$Q_{i+1} = Q_i + \frac{u_i u_i^{\mathrm{T}}}{u_i^{\mathrm{T}} y_i} - \frac{Q_i y_i y_i^{\mathrm{T}} Q_i^{\mathrm{T}}}{y_i^{\mathrm{T}} Q_i y_i}.$$

The above recurrence relationships provide a continuous revision of the model parameters with minimal computational effort. The parameters are updated in a single-step operation at each successive time interval by re-evaluating the gradient so as to incorporate the latest measurements. The resulting model will be the best, in a least-squares sense, that can be fitted to the data w_t up to the current time N. If now the sample autocorrelation function of the residual error between the model and actual data at each point given by

$$a_t = \theta^{-1}(B)\Theta^{-1}(B)\phi(B)\Phi(B) \tag{12.20}$$

is computed then $r_k(a_t)$ can be used as an indication of the success of the modelling.

5 DEMAND PREDICTION

Having established a model of the time series w_t derived from the demand data over the period $t = 1, \ldots, N$ in the form of equation (12.13). this model may now be used for prediction by expanding it forward in time $t = N+1, \ldots, N+k$ with the assumption that the white noise error function a_t is zero for $t > n$. Thus

$$F D \mathbf{w}' = R T \mathbf{a}' \tag{12.21}$$

where

$$\mathbf{w}'^{\mathrm{T}} = [\mathbf{w}^{\mathrm{T}} : w_{N+1}, \ldots, w_{N+k}]$$

$$\mathbf{a}'^{\mathrm{T}} = [\mathbf{a}^{\mathrm{T}} : 0, \ldots, 0]$$

and all the matrices are now $(N+k) \times (N+k)$

Introducing new variables

$$\mathbf{a}'' = T\mathbf{a}' \tag{12.22}$$

$$\mathbf{a}^* = R\mathbf{a}'' \tag{12.23}$$

it is easy to see that computations of the values $a_t^*, t = N+1, \ldots, N+k$, requires only a_t'', $t = (N+1-sQ), \ldots, N+k$, which in turn requires only a_t', $t = (N+1-sQ-q), \ldots, N+k$. Similarly by introducing

$$\mathbf{w}'' = D\mathbf{w}' \tag{12.24}$$

$$\mathbf{w}^* = F\mathbf{w}'' \tag{12.25}$$

it can be shown that to solve (12.24) for w_t', $t = N+1, \ldots, N+k$, requires w_t'', $t = N+1, \ldots, N+k$. These in turn can be found from equation (12.25), given w_t'', $t = (N+1-sP), \ldots, N$, which are calculated from (12.25) using known values w_t', $t = (N+1-sP-p), \ldots, N$.

The probability limits for the forecast can be determined by re-writing equations (12.2) and (12.3) as

$$z_t = (1 + \Psi_1 B + \Psi_2 B^2 + \ldots) a_t. \tag{12.26}$$

The variance of the forecast errors $[z_{t+1} - z_t(l)]$, where $z_t(l)$ represents predicted transformed data, is given by

$$V(z_{t+l} \mid a_t, a_{t-1}, \ldots) = \left(1 + \sum_{j=1}^{l-1} \Psi_j^2\right) \sigma_a^2 \tag{12.27}$$

where

$$\sigma_a^2 = \frac{1}{n} \sum_{t}^{n} a_t^2.$$

The 95% probability limits on the forecast are consequently given by

$$z_t(1) \pm 2\left(1 + \sum_{j=1}^{l-1} \Psi_j^2\right) \sigma_a^2. \tag{12.28}$$

6 COMPUTATIONAL RESULTS

A software package has been developed in the Department of Engineering at the University of Durham to implement the above ARMA modelling and prediction technique together with on-line parameter tracking. The software includes colour graphical display of logged and predicted demand, together with an error analysis enabling an interactive selection of the model structure before the prediction is run. The method has been successfully applied both to power and water demand data giving prediction errors of a few percent over an extended time period.

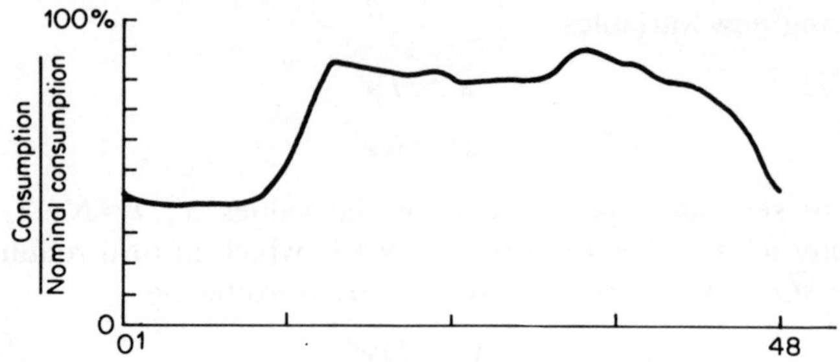

FIGURE 12.1 Daily consumption pattern

The performance of the above technique has been investigated for half-hourly water demand data shown in Figure 12.1 and an appropriate model structure found to be

$$(1+\phi_1 B)(1+\Phi_1 B^{48})w_t = (1+\theta_1 B)(1+\Theta_1 B^{48})a_t \tag{12.29}$$

which gives an almost exact fit to the data.

The autocorrelation function of the residuals between the actual data values and the values calculated from the model is given in Figure 12.2. Although expansion of the model (12.29) of second-order nonseasonal components improves the autocorrelation function, it has been found that it has only a marginal effect on prediction accuracy. Figure 12.3 shows that the prediction error over short times is of order 1.0% with an increase to approximately 5.0% for longer term forecasts. This significant worsening of the longer term prediction error is caused by the absense of a model component reflecting weekly periodicity which becomes more important as the prediction time expands. However, an improvement in longer term

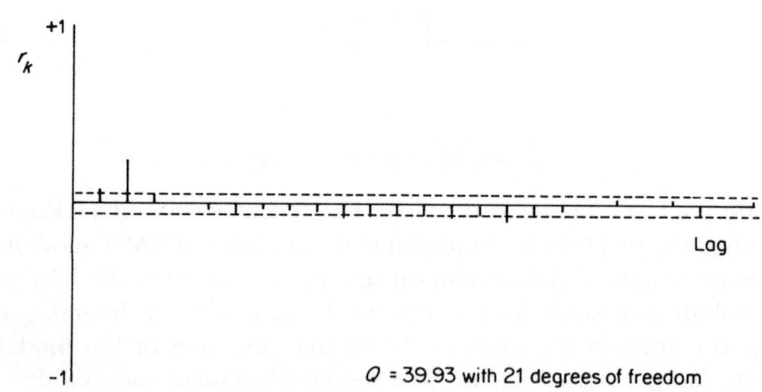

FIGURE 12.2 Autocorrelation function of residuals

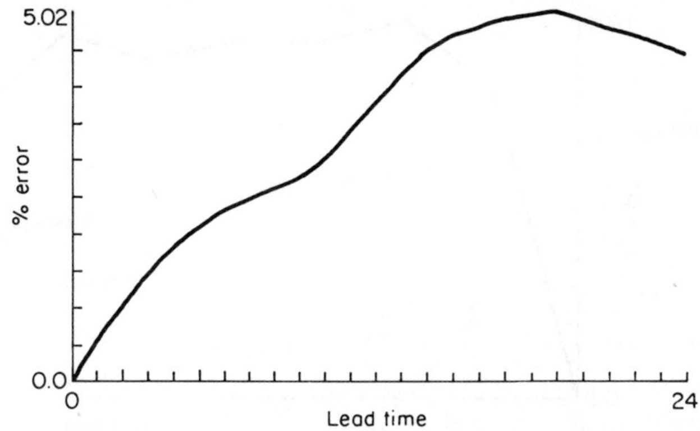

FIGURE 12.3 Mean absolute prediction error for up to 24 steps ahead forecast after 48 prediction steps

accuracy can be achieved by introducing an additional model, based on average daily consumption data, which would facilitate predictions for several days ahead.

The daily demand model would then have the form

$$(1+\phi_1 B)w_t = (1+\theta_1 B)(1+\Theta_1 B^7)a_t \tag{12.30}$$

and produces the autocorrelation function of residuals shown in Figure 12.4. The prediction error is now reduced to approximately 1.7% (Figure 12.5). Introduction of a hierarchy of models according to their time scales allows a simple structure to be retained and consequently enhances the computational efficiency of the method.

The significance of the tracking of model parameters has also been studied. Figures 12.6–12.9 illustrate the variation of parameters of the

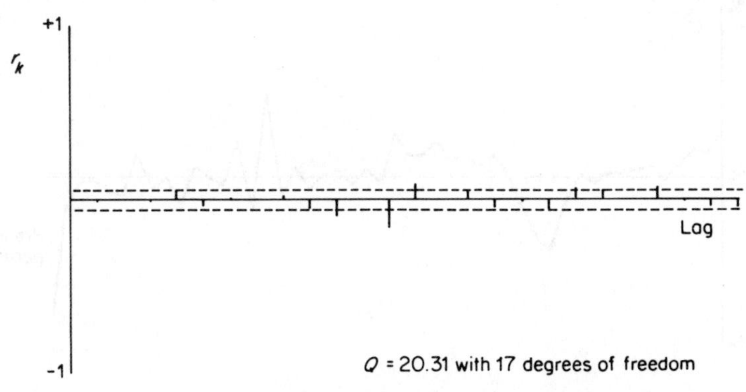

FIGURE 12.4 Autocorrelation function of residuals

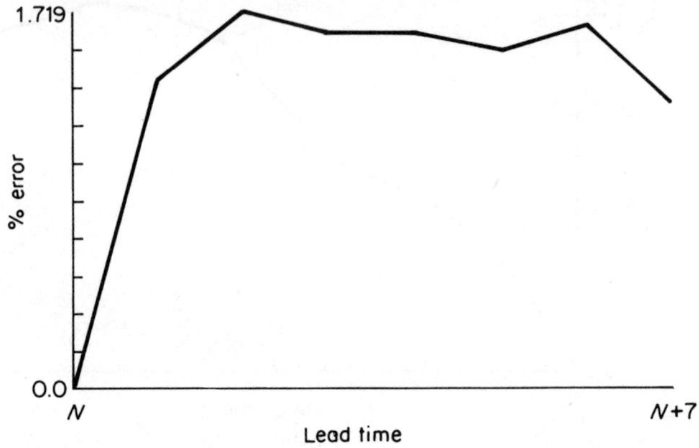

FIGURE 12.5 Mean absolute prediction error for up to 7 steps ahead forecast after 14 prediction steps

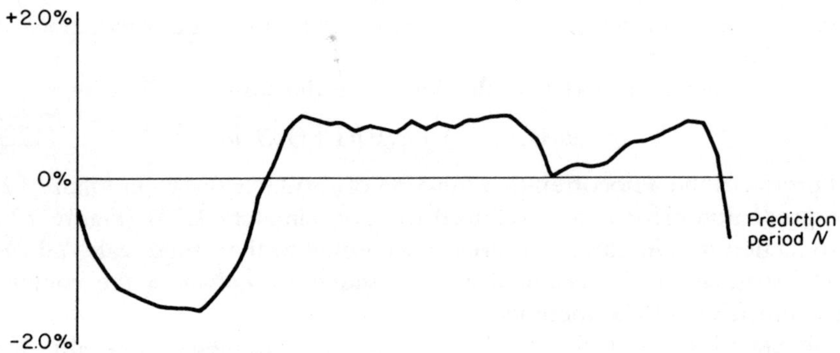

FIGURE 12.6 Percentage change in ϕ_1 from its mean value of 0.966

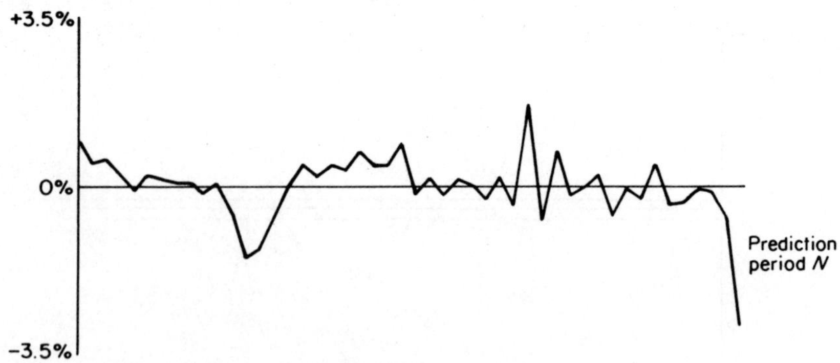

FIGURE 12.7 Percentage change in Φ_1 from its mean value of 0.981

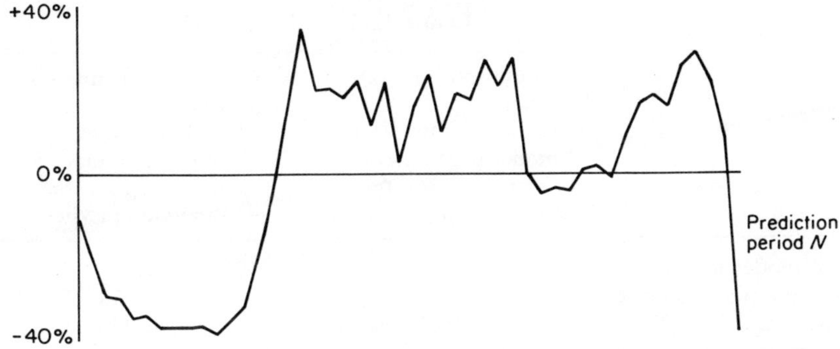

FIGURE 12.8 Percentage change in θ_1 from its mean value of 0.564

model (12.29) during a 24-hour period. As expected, the seasonal autoregressive parameter Φ_1 shows only some random fluctuations around its mean value (Figure 12.7) since each new data point remains in agreement with the periodicity of the demand. The parameters ϕ_1, θ_1, and Θ_1 are modified in line with the changes of the demand giving a good justification for the adaptive procedure. As can be seen from Figures 12.8 and 12.9 the changes of magnitude of the moving average parameters θ_1 and Θ_1 are particularly high and alter by as much as +35% to −39% and +42% to −28% respectively of the mean value of the parameters. This lends support to the conclusion that the adaptive structure of the model is highly desirable.

Computation times obtained for a FORTRAN 77 implementation of the algorithm using a 32-bit minicomputer (Perkin–Elmer 3220) are presented in Table 12.1.

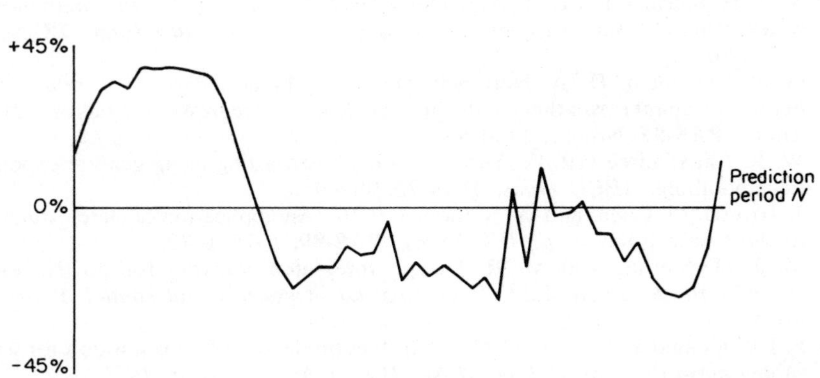

FIGURE 12.9 Percentage change in Θ_1 from its mean value of −0.377

TABLE 12.1

	Computation time (s) Model 1 –336 data points –4 model parameters –48 time-step season –24 time-atep prediction	Computation time (s) Model 2 –100 data points –3 model parameters –7 time–step season –7 time-step prediction
Initial model fitting	8.0	1.6
Model parameters update	0.720	0.150
Prediction	0.085	0.030

7 CONCLUSIONS

A technique for the modelling of time series has been presented and shown to be of a form suitable for on-line prediction of water demand but the method has also been found to be suitable for electricity demand prediction. A novel application of the Fletcher–Powell variable metric algorithm for function minimization in the Newton–Raphson method has been shown to result in a unity rank updating formula which enables the tracking of variation in the model parameters. The significance of the adaptive structure of the algorithm has been illustrated above and found to be a highly desirable feature for an on-line computer-based demand prediction scheme.

REFERENCES

1. E. D. Farmer (1963). 'A method of prediction for nonstationary processes and its application to the problem of load estimation.' *Proc. IFAC Congress, Basle, 1963.*
2. M. J. H. Sterling and D. J. Antcliffe (1974). 'A technique for the prediction of water demand from past consumption data.' *J. Inst. of Water Engs.*, **28,** No. 8, 413–420.
3. G. T. Heineman, D. A. Nordman, and E. C. Plant (1966). 'The relationship between summer weather and summer load—A regression analysis.' *IEEE Trans.*, **PAS-85,** No. 11, 1144–1154.
4. W. R. Christiansee (1971). 'Short term load forecasting using general exponential smoothing.' *IEEE Trans.*, **PAS-90,** 909–910.
5. J. Toyoda, M. Chen, and W. S. Inoue (1970). 'An application of state estimation to short term forecasting.' *IEE Trans.*, **PAS-89,** 1678–1682.
6. M. J. H. Sterling and M. R. Irving, 'Integrated software for power system control.' In M. Singh (Ed.), *Encyclopaedia of systems and control,* Pergamon Press.
7. F. Fallside and P. F. Perry (1975). 'On-line prediction of consumption for water supply network control.' *Proc. IFAC, Boston, Mass., August 1975.*
8. R. DeMoyer and L. B. Horwitz (1975). *A system approach to water distribution modelling and control,* Lexington Books, D. C. Heath & Co.

9. G. E. P. Box and G. M. Jenkins (1970). *Time series analysis: Forecasting and control*, Holden-Day, San Francisco.
10. M. J. H. Sterling (1978). *Power systems control*, Peter Peregrinus, London.
11. W. Gilchrist (1976). *Statistical forecasting*, Wiley, London.
12. R. Fletcher and M. J. D. Powell (1963). 'A rapidly convergent descent method for minimisation.' *Comput. J.*, **6**, 163–168.

Author Index

Subject Index